CRC Handbook
of
Oligosaccharides

Volume II
Trisaccharides

Authors

András Lipták, D.Sc.
Professor
Institute of Biochemistry
Lajos Kossuth University
Debrecen, Hungary

Zoltán Szurmai, Ph.D.
Scientific Fellow
Institute of Biochemistry
Lajos Kossuth University
Debrecen, Hungary

Péter Fügedi, Ph.D.
Scientific Fellow
Institute of Biochemistry
Lajos Kossuth University
Debrecen, Hungary

János Harangi, Ph.D.
Scientific Fellow
Institute of Biochemistry
Lajos Kossuth University
Debrecen, Hungary

CRC Press
Taylor & Francis Group
Boca Raton London New York

CRC Press is an imprint of the
Taylor & Francis Group, an **informa** business

First published 1991 by CRC Press
Taylor & Francis Group
6000 Broken Sound Parkway NW, Suite 300
Boca Raton, FL 33487-2742

Reissued 2018 by CRC Press

© 1991 by Taylor & Francis
CRC Press is an imprint of Taylor & Francis Group, an Informa business

No claim to original U.S. Government works

Publisher's Note
The publisher has gone to great lengths to ensure the quality of this reprint but points out that some imperfections in the original copies may be apparent.

Disclaimer
The publisher has made every effort to trace copyright holders and welcomes correspondence from those they have been unable to contact.

ISBN 13: 978-1-138-10534-8 (hbk)
ISBN 13: 978-1-138-55838-0 (pbk)
ISBN 13: 978-1-315-15082-6 (ebk)

Visit the Taylor & Francis Web site at http://www.taylorandfrancis.com and the
CRC Press Web site at http://www.crcpress.com

INTRODUCTION

The number of free oligosaccharides occurring in nature is relatively low, but their derivatives in the form of, for example, plant glycosides, antibiotics, glycolipids, and mainly glycoproteins are indeed widespread. Especially in the case of the last two groups, to clear up their biological roles, intense research has begun not only in the circles of biochemists and immunologists but also among the synthetic carbohydrate chemists.

The improvement in the efficiency of isolation techniques and high performance spectroscopic methods has enabled unprecedented development in the area of oligosaccharides. The extreme variety of structures and in some cases the miniscule amounts of isolated materials involved have presented new challenges for synthetic chemists; new synthetic methods had to be developed to ensure sufficient quantities for biological investigations and to enable the production of varied structures. Extraordinary progress has been achieved in two areas of synthesis. First, with the aid of new blocking strategies a wide range of partially protected mono- and oligosaccharides have become available. Secondly, with the exploitation of the more detailed mechanism of glycosylation reactions, new methods appeared which were suitable for the production of 1,2-cis and 1,2-trans glycosides. As a result, in the past 10 years an extremely wide range of oligosaccharide syntheses could be achieved.

Although during the past decade a number of excellent reviews have been published to assist in the survey of the newest developments in the field, there has been no book that would provide a survey of the structures of the oligosaccharides synthesized to date and the details of the applied methods. We felt, on the basis of our own experience, that such a book would represent a great help to the carbohydrate chemist and at the same time to all scientists who deal in some ways with oligosaccharides.

Our collection is meant to meet the needs, first of all, of those dealing in synthetic carbohydrate chemistry and beyond merely listing the syntheses, also showing the route of the synthesis. Thus, not only are those oligosaccharides presented which were prepared in their free form, but also those produced in a protected form. In the planned series of three volumes can be found all of the oligosaccharides synthesized between 1960 and 1986. We present in tabular form the aglycon, the glycosyl donor, the reaction conditions applied (solvent, promoter, temperature), and the structure of the isolated product. In the case of disaccharides the names of the reactants and the products are given, while in the case of the trisaccharides and higher oligomers schematic figures provide quick and easy information concerning the entire synthesis.

THE AUTHORS

András Lipták, Ph.D., D.Sci., is Professor of Biochemistry in the Department of Biology at Kossuth Lajos University, Debrecen, Hungary. He received his M.Sc. degree from the same university in 1961, graduating with highest honors. In 1968 he received his Ph.D. in organic chemistry and in 1983 he was honored by the Hungarian Academy of Sciences as D.Sc.

Professor Lipták held an Alexander von Humboldt Fellowship in Munich in 1971-1972 and he spent nearly 2 years at the National Institutes of Health (Bethesda, Maryland). He is an author of over 110 articles and has presented his scientific results at numerous international meetings. He is the member of the Editorial Board of the *Journal of Carbohydrate Chemistry*. His main research interest covers the selective protecting of carbohydrates and synthesis of complex oligosaccharides. In 1989 the Hungarian Academy of Sciences awarded him with the Zemplén Géza Prize.

Zoltán Szurmai, Ph.D., is a scientific fellow at the Institute of Biochemistry, Kossuth Lajos University in Debrecen, Hungary. He graduated from Kossuth Lajos University in 1977 as a chemist. Dr. Szurmai received his Ph.D. degree from the same university in 1982. He then completed a 1-year fellowship at the Gerontology Research Center, National Institutes of Health, National Institute on Aging, in Baltimore, Maryland. He worked for Ruhr University in 1988 and 1989 for a short time in West Germany. His major research interest is the chemical syntheses of oligosaccharides. He has 17 articles published in international journals.

Péter Fügedi, Ph.D., is currently a visiting scientist at Glycomed Inc. in Alameda, California. He received his M.Sc. degree in chemistry at Kossuth Lajos University, Debrecen, Hungary in 1975, and his Ph.D. degree in 1978 from the same university. He got his habilitation from the Hungarian Academy of Sciences in 1989. Being affiliated at the Institute of Biochemistry of Lajos Kossuth University, he had postdoctoral experience at the Laboratory of Structural Biochemistry in Orleans, France in 1978-1979, and was a visiting scientist at the Department of Organic Chemistry, Arrhenius Laboratory, University of Stockholm, Sweden in 1984-1985. He is an author of over 30 articles and presented his results at numerous international meetings. His major research interest is synthetic carbohydrate chemistry, with emphasis on glycosylation methods and the synthesis of oligosaccharides.

János Harangi, Ph.D., is a scientific fellow at the Institute of Biochemistry, Kossuth Lajos University, Debrecen, Hungary. He graduated from Kossuth Lajos University in 1974 as a chemist. In 1980 he received his Ph.D. in biochemistry from the same university. He worked at Munich University and Ruhr University in West Germany several times. His main research interest is the structure investigation of sugar derivatives by nuclear magnetic resonance spectroscopy and separation techniques in carbohydrate chemistry. He has 30 articles published in international journals. He is now working for Hewlett-Packard Vienna as a sales representative and customer trainer.

ADVISORY BOARD

HOW TO USE THE BOOK

Handbook of Oligosaccharides, Volume II contains all the chemical syntheses of 358 individual trisaccharides published between 1960—1986 in the following sequence of monosaccharide units at the reducing end:

D-Riboses
D-Arabinose
L-Arabinoses
D-Xyloses
D-Glucoses
D-Glucuronic acids
2-Amino-2-deoxy-D-glucoses
D-Mannoses
2-Amino-2-deoxy-D-mannose
L-Rhamnoses
D-Galactoses
2-Amino-2-deoxy-D-galactoses
L-Fucoses
D-Fructoses
KDO
Deoxy sugars

In the hierarchy of the monosaccharide building blocks the following order is used: Ribose, Arabinose, Xylose, Lyxose, Allose, Altrose, Glucose, Mannose, Gulose, Idose, Galactose, Talose, Fructose, KDO, Neuraminic Acid, Deoxy Sugars.

The D-sugars precede the L-sugars. The uronic acids follow immediately after the parent sugars. Some deoxy- and aminodeoxy-sugars such as 2-amino-2-deoxy-D-glucose, -D-mannose, and -D-galactose can be found immediately after uronic acids. Among the 6-deoxy-sugars, the rhamnose (6-deoxy-mannose) and fucose (6-deoxy-galactose) have the same privilege. All other amino-sugars, deoxy-sugars, and deoxy-halogenosugars are itemized after the neuraminic acid. The branched chain sugars follow these derivatives.

Concerning the ring size, the furanosides precede the pyranosides. The α-anomers precede the β-anomers.

The sequence of the bond types is the following: (1→1); (1→2); (1→3); (1→4); i.e., instead of the full name of trisaccharides the abbreviated name is given, e.g., α-D-Manp-(1→4)-α-L-Rhap-(1→3)-D-Gal instead of O-(α-D-Mannopyranosyl)-(1→4)-O-(α-L-rhamnopyranosyl)-(1→3)-D-galactose. Linear trisaccharides precede branched ones when they have the same sugar at the reducing end. If a trisaccharide is not listed it was most probably not prepared synthetically in that period.

Under an individual trisaccharide entry you can find the following information: the abbreviated name of the trisaccharide; physical data of the free trisaccharide (m.p. α_D value) if it is prepared. In some cases if the most important glycosides of the trisaccharide are known, one can find symbols of these; e.g. β-p-(1→OMe) means that the β-methyl glycoside of the trisaccharide was prepared and the reducing end was in pyranoside form; physical data of the important glycosides; also one can get information about the glycosylation reactions: there are schematic figures of the aglycons on the far left and the glycosyl donors left of center; in the middle, one can see the reaction conditions (catalyst, solvent). Different reaction conditions are numbered with roman figures.

We used abbreviations for catalysts and solvents (see list of abbreviations). Schematic figures of the products are located on the far right. Underneath the figures one can find data of the products (e.g. yield, m.p., α_D value). C-13 indicates that ^{13}C-NMR data are available in the original paper. We used abbreviations for the protecting groups in schematic figures (see list of abbreviations).

ABBREVIATIONS

Sugars

All	=	Allose
Alt	=	Altrose
Ara	=	Arabinose
Colp	=	3,6-Dideoxy-L-*xylo*-hexopyranosyl
Fuc	=	Fucose
FucNAc	=	2-Acetamido-2-deoxy-fucose (2-acetamido-2,6-dideoxy-galactose)
Fuc4NAc	=	4-Acetamido-4-deoxy-fucose (4-acetamido-4,6-dideoxy-galactose)
Fru	=	Fructose
Gal	=	Galactose
GalA	=	Galacturonic acid
GalN	=	2-Amino-2-deoxy-galactose
GalNAc	=	2-Acetamido-2-deoxy-galactose
Glc	=	Glucose
GlcA	=	Glucuronic acid
GlcN	=	2-Amino-2-deoxy-glucose
GlcNAc	=	2-Acetamido-2-deoxy-glucose
Hex	=	Hexose
Hep	=	Heptose
Idop	=	Idopyranosyl
KDO	=	3-Deoxy-D-*manno*-2-octulopyranosic acid
Lyxp	=	Lyxopyranosyl
Man	=	Mannose
ManN	=	2-Amino-2-deoxy-mannose
ManNAc	=	2-Acetamido-2-deoxy-mannose
ManNAcA	=	2-Acetamido-2-deoxy-mannuronic acid
Neu5Ac	=	N-Acetyl-neuraminic acid
Rha	=	Rhamnose
Rib	=	Ribose
Tyvp	=	Tyvelopyranosyl (3,6-dideoxy-D-*arabino*-hexopyranosyl)
Xyl	=	Xylose

Protecting groups

Ac	=	Acetyl
All	=	Allyl
Bz	=	Benzoyl
Bzl	=	Benzyl
Bu	=	Butyl
tBu	=	*tert*-Butyl
Bue	=	Butenyl
tBuMe$_2$Si	=	*tert*-Butyldimethylsilyl
Bu$_3$Sn	=	Tributylstannylene
Cbz	=	Carbobenzyloxy

C(Me)$_2$	=	Isopropylidene
Dcp	=	2,3-Diphenyl-2-cyclopropenyl
DNP	=	Dinitrophenyl
ECO	=	8-Ethoxycarbonyloctyl
Et	=	Ethyl
cHex	=	*cyclo*-Hexylidene
MBA	=	Monobromoacetyl
MBz	=	*p*-Methoxybenzoyl
MCA	=	Monochloroacetyl
MCO	=	8-Methoxycarbonyloctyl
Me	=	Methyl
pMeOPhCH	=	*p*-Methoxybenzylidene
NBz	=	*p*-Nitrobenzoyl
oNBz	=	*o*-Nitrobenzoyl
Ph	=	Phenyl
PhCH	=	Benzylidene
Phth	=	Phthalimido
PNP	=	*p*-Nitrophenyl
Pr	=	Propyl
Pre	=	Prenyl
Sp	=	Spacer
TCA	=	Trichloroacetyl
TCE	=	2,2,2-Trichloroethyl
Ts	=	*p*-Toluenesulfonyl
Tr	=	Triphenylmethyl

Reagents and solvents

Ac$_2$O	=	Acetic anhydride
AcOH	=	Acetic acid
AgOTf	=	Silver trifluoromethanesulfonate
AgSi	=	Silver silicate
AllBr	=	Allyl bromide
Bu$_4$NBr	=	Tetrabutylammonium bromide
(Bu$_3$Sn)$_2$O	=	Tributyltin oxide
BzlBr	=	Benzyl bromide
s-Coll	=	Collidine (2,4,6-trimethylpyridine)
DBU	=	1,8-Diazabicyclo[5.4.0]undec-7ene
DMA	=	N,N-Dimethylacetamide
DMF	=	N,N-Dimethylformamide
DMSO	=	Dimethyl sulfoxide
DpcClO$_4$	=	2,3-Diphenyl-2-cyclopropen-1-ylium perchlorate
Et$_3$N	=	Triethylamine
Et$_4$NBr	=	Tetraethylammonium bromide
Et$_4$NCl	=	Tetraethylammonium chloride
EtOAc	=	Ethyl acetate
Et$_2$O	=	Diethyl ether
EtOH	=	Ethyl alcohol
Hg(OAc)$_2$	=	Mercuric acetate
LutClO$_4$	=	2,6-Dimethylpyridinium perchlorate
MeCN	=	Acetonitrile
MeNO$_2$	=	Nitromethane

MeOH	=	Methyl alcohol
MeOTf	=	Methyl trifluoromethanesulfonate
Me$_3$SiBr	=	Trimethylsilyl bromide (Bromotrimethylsilane)
NaOAc	=	Sodium acetate
NaOMe	=	Sodium methoxide
pNBCl	=	*p*-Nitrobenzenesulfonyl chloride
NIS	=	N-Iodosuccinimide
PdC	=	Palladium on carbon
(iPr)$_2$EtN	=	N,N-Diisopropylethylamine
Py	=	Pyridine
TMSOTf	=	Trimethylsilyl trifluoromethanesulfonate
TMU	=	1,1,3,3-Tetramethylurea
TrBF$_4$	=	Triphenylmethylium tetrafluoroborate
TrClO$_4$	=	Triphenylmethylium perchlorate
pTSA	=	*p*-Toluenesulfonic acid

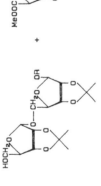

α-D-Ribf-(1→5)-α-D-Ribf-(1→5)-D-Rib

Hygroscopic solid: [α]_548 +122.6 (methanol): Ref.: 1

Ag₂O
chloroform

R=Me: 77%; oil; [α]_548 +9.2 (chloroform): Ref.: 1,2
R=Bzl: 76.5%; oil; [α]_548 -13.1 (chloroform): Ref.: 1

β-D-RibfA-(1→5)-α-D-Ribf-(1→5)-D-Rib

Ag₂O
chloroform

R=Me: 8%; oil: Ref.: 1
R=Bzl: 8.5%; oil: Ref.: 1

α-KDO-(2→2)-β-D-Ribf-(1→2)-D-Rib

β-f-(1-OMe) Na-salt: colourless glass: [α]$_D$ +16.1 (water): Ref.: 3

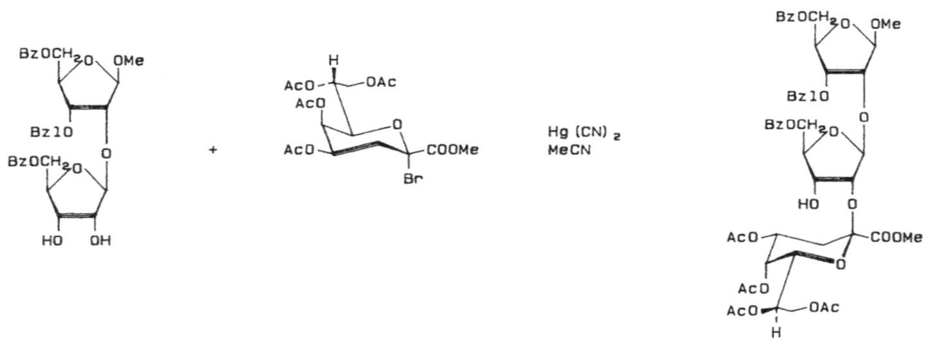

Hg(CN)$_2$
MeCN

25%: [α]$_D$ +92.7 (chloroform): Ref.: 3

α-D-Ribf-(1→2)
α-KDO-(2→3) ——— D-Rib

Ref.: 3

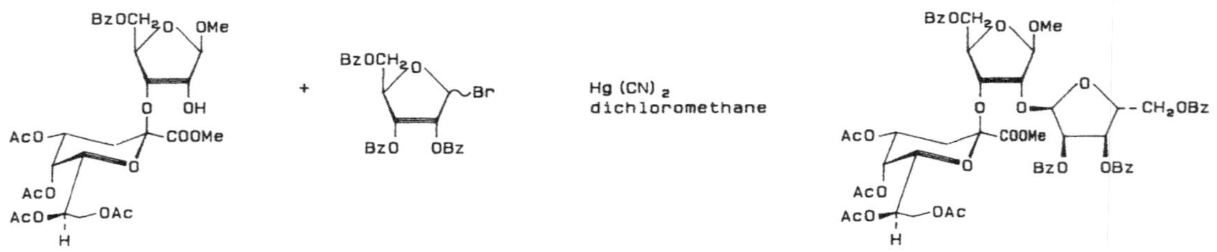

Hg(CN)$_2$
dichloromethane

[α]$_D$ +91.1 (chloroform)

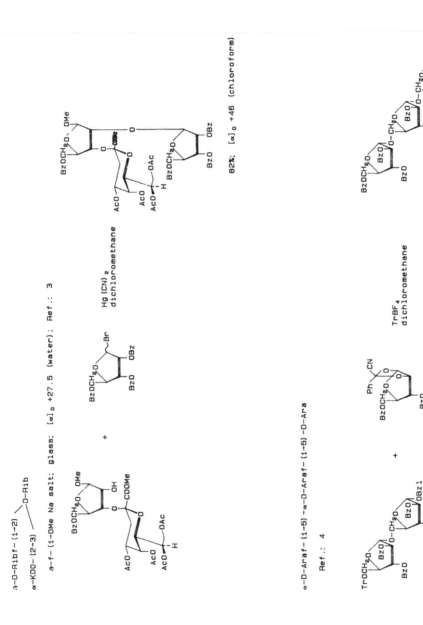

β-D-Ribf-(1→2)
                    ⟍
                      D-Rib
α-KDO-(2→3)   ⟋

β-f-(1-OMe Na salt; glass; [α]$_D$ +27.5 (water); Ref.: 3

Hg(CN)$_2$
dichloromethane

82%; [α]$_D$ +46 (chloroform)

α-D-Araf-(1→5)-α-D-Araf-(1→5)-D-Ara

Ref.: 4

TrBF$_4$
dichloromethane

35%; [α]$_D$ +11.3 (chloroform)

α-L-Rhap-(1→2)
\   L-Ara
β-D-Glcp-(1→3)

amorphous; $[\alpha]_D$ 0 (water); C-13; Ref.: 5

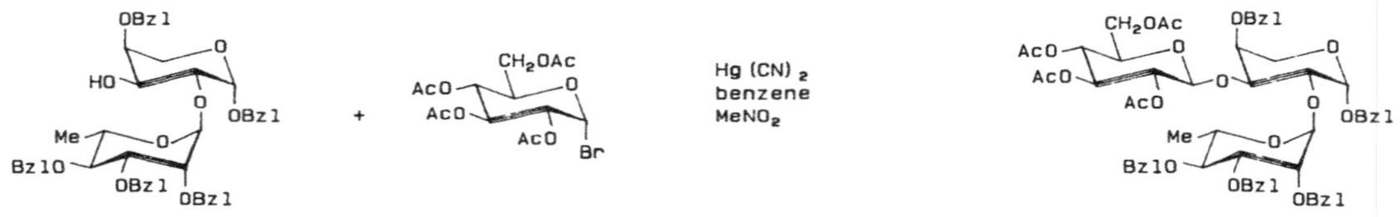

+

Hg(CN)$_2$
benzene
MeNO$_2$

74%; $[\alpha]_D$ +45 (chloroform)

α-L-Rhap-(1→2)
\   L-Ara
β-D-Glcp-(1→4)

amorphous; $[\alpha]_D$ +19 (water); C-13; Ref.: 5

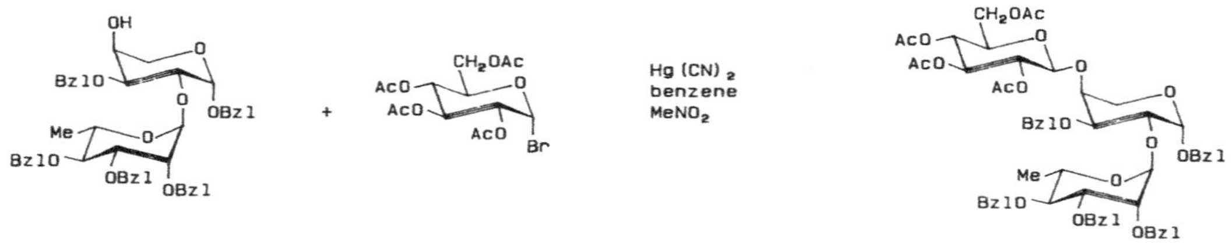

+

Hg(CN)$_2$
benzene
MeNO$_2$

53%; $[\alpha]_D$ +42 (chloroform)

ß-D-Xylp-(1→4)-α-D-Xylp-(1→4)-D-Xyl

Ref.: 6

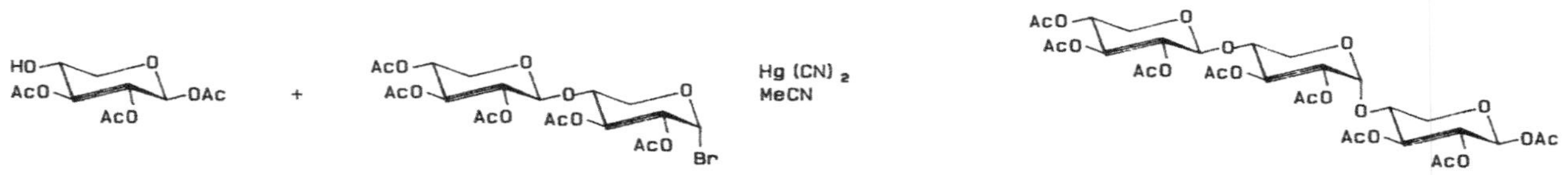

9.4%; [α]_D -13.5 (chloroform)

α-D-Xylp-(1→4)-ß-D-Xylp-(1→2)-D-Xyl

ß-p-(1→OMe); m.p. 135-138 °C; [α]_D 13.9 (water); C-13; Ref.: 7

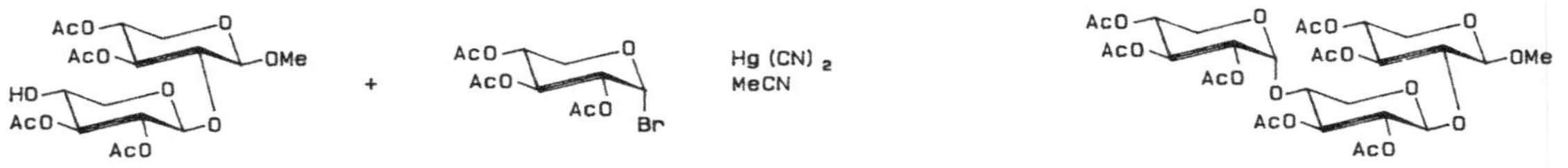

23%; m.p. 160-162°C; [α]_D +9.2 (chloroform)

β-D-Xylp-(1→4)-β-D-Xylp-(1→2)-D-Xyl

β-D-(1→OMe): m.p. 158-159.5°C; [α]$_D$ -73.2 (water); C-13; Ref.: 7

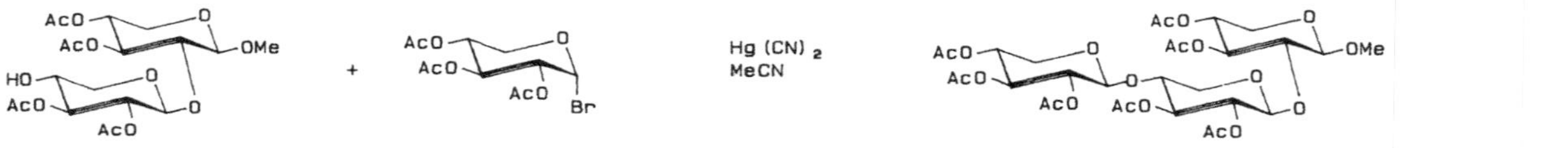

Hg(CN)$_2$
MeCN

68%; m.p. 152-154°C; [α]$_D$ -80.2 (chloroform)

α-D-Xylp-(1→4)-α-D-Xylp-(1→3)-D-Xyl

β-D-(1→OMe): m.p. 211-212.5°C; [α]$_D$ +18.3 (water); C-13; Ref.: 7

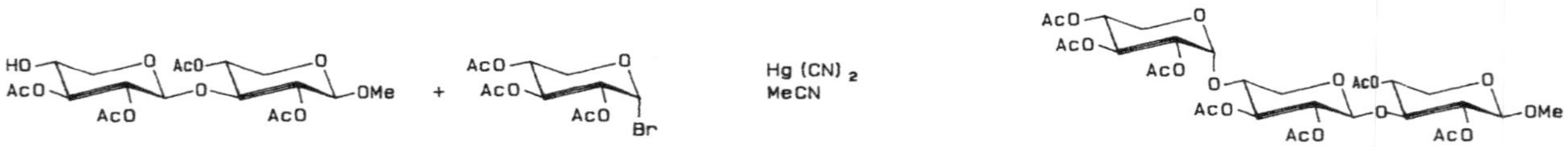

Hg(CN)$_2$
MeCN

23%; m.p. 189-190°C; [α]$_D$ -13.2 (chloroform)

β-D-Xylp-(1→3)-β-D-Xylp-(1→3)-D-Xyl

Ref.: 8

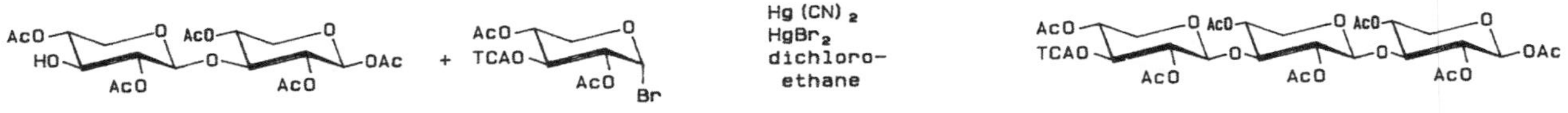

Hg(CN)₂
HgBr₂
dichloro-
ethane

m.p. 190 °C; [α]ᴅ -52 (chloroform)

β-D-Xylp-(1→4)-β-D-Xylp-(1→3)-D-Xyl

β-D-(1→OMe); m.p. 178-180 and 199-200 °C; [α]ᴅ -77.4; C-13; Ref.: 7

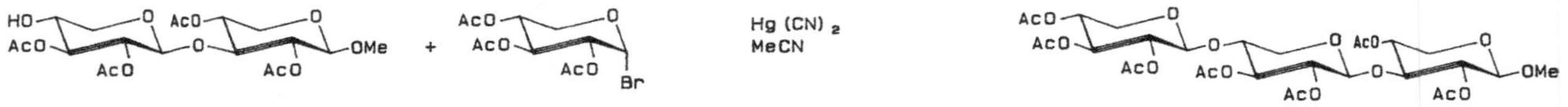

Hg(CN)₂
MeCN

70%; m.p. 178-179 °C; [α]ᴅ -101.2 (chloroform)

α-L-Araf-(1→3)-β-D-Xylp-(1→4)-D-Xyl

[α]$_D$ -72 (water); Ref.: 9

β-D-(1→OMe): m.p. 146-147°C; [α]$_D$ -121 (water); C-13; Ref.: 10

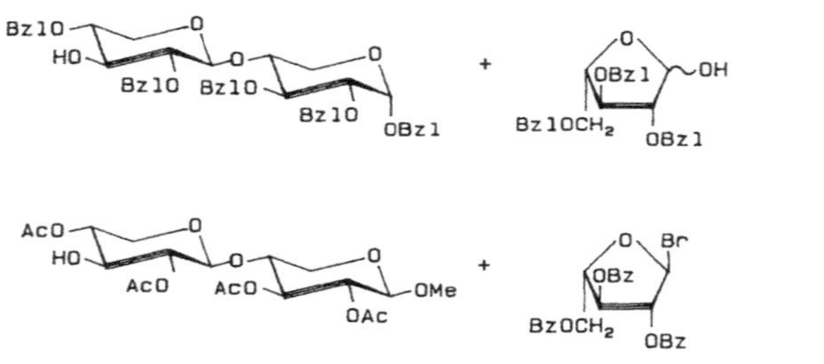

AgOTf
DBU
NsCl
dichloro-
methane
Ref.: 9

Hg(CN)$_2$
benzene
Ref.: 10

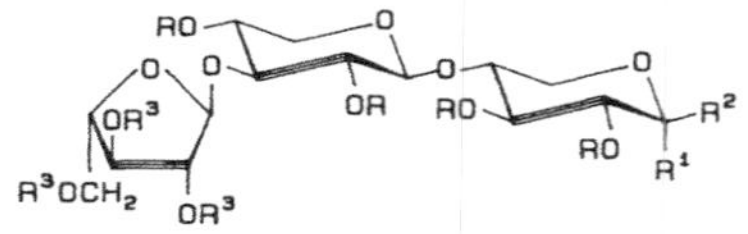

R^1=OBzl; R^2=H; R=R^3=Bzl; [α]$_D$ +14 (chloroform); Ref.: 9

R^2=OMe; R^1=H; R=Ac; R^3=Bz; 91.1%; [α]$_D$ -48 (chloroform); C-13; Ref.: 10

α-D-Xylp-(1→3)-α-D-Xylp-(1→4)-D-Xyl

β-D-(1→OMe): m.p. 218-223 °C; [α]$_D$ +30.5 (water); Ref.: 11

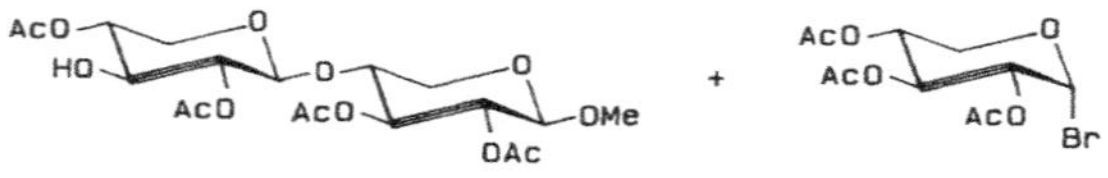

Hg(CN)$_2$
MeCN

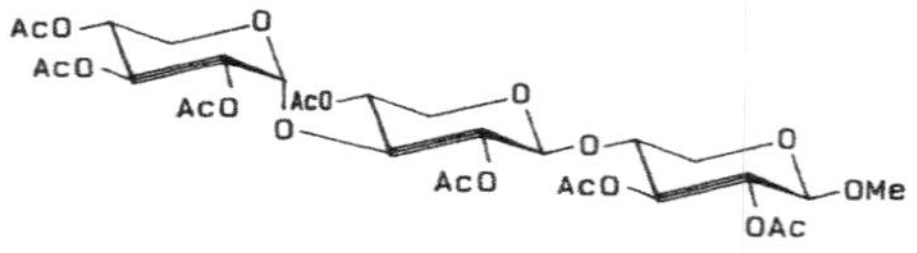

18%; m.p. 120-127°C; [α]$_D$ -4 (chloroform); Ref.: 11

α-D-Xylp-(1→4)-β-D-Xylp-(1→4)-D-Xyl

β-D-(1-OMe) acetate: [α]$_D$ -16 (chloroform); C-13; Ref.: 12

α-acetate: m.p. 113-119 °C; [α]$_D$ -0.6 (chloroform); C-13; Ref.: 13

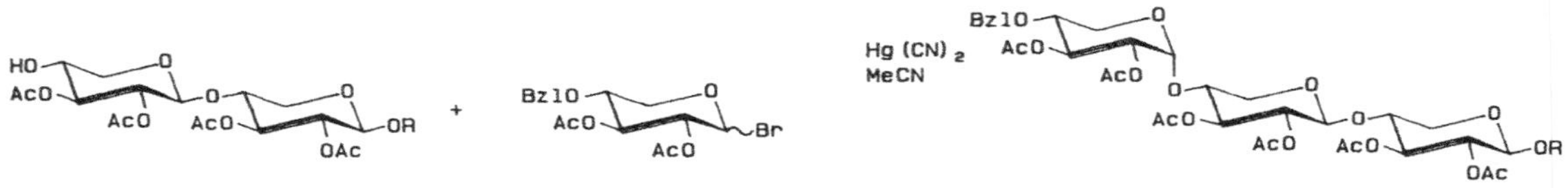

Hg(CN)$_2$
MeCN

R=Me: 27%; [α]$_D$ -18 (chloroform); C-13; Ref.: 12

R=AC: 36%; [α]$_D$ -4.4 (chloroform); C-13; Ref.: 13

β-D-Xylp-(1→2)-β-D-Xylp-(1→4)-D-Xyl

β-D-(1-OMe): m.p. 202-203 °C; [α]$_D$ -80 (water); Ref.: 14

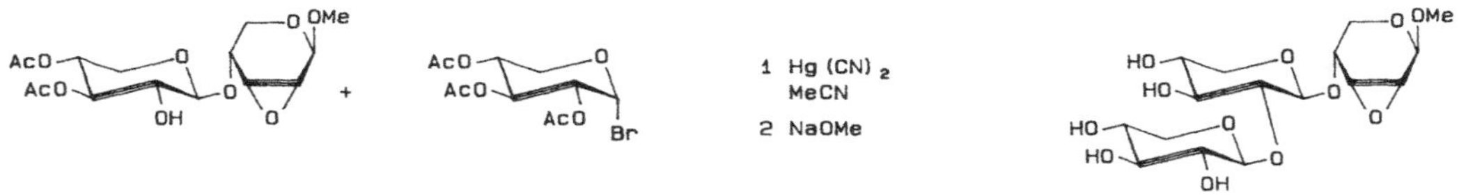

1 Hg(CN)$_2$
  MeCN
2 NaOMe

44.1%; m.p. 222-224 °C; [α]$_D$ -47 (water); Ref.: 14

β-D-Xylp-(1→4)-β-D-Xylp-(1→4)-D-Xyl

m.p. 217-219 °C; [α]$_D$ -47.7 (water); C-13; Ref.: 13

m.p. 222-224 °C (EtOH); m.p. 217-219 °C (MeOH); [α]$_D$ -47.7 (water); C-13; Ref.: 6

β-p-(1-OMe); m.p. 186-190 °C; C-13; Ref.: 12

β-p-(1-OMe); m.p. 183-186 °C; [α]$_D$ -80.5 (water); Ref.: 15

β-p-(1-OMe); m.p. 190-191 °C (several recr. from MeOH); Ref.: 15

I Hg(CN)$_2$ MeCN R=Me: Ref.: 12, 16 R=Ac: Ref.: 13, 16

I Hg(CN)$_2$ MeCN R=Me: Ref.: 15 R=Ac: Ref.: 6

II 1 Hg(CN)$_2$ MeCN 2 NaOMe

I R=Me; R'=Bzl; 52.4%; m.p. 155-157 °C; [α]$_D$ -98 (chloroform); C-13; Ref.: 12, 16

R=Me; R'=Ac; 41%; m.p. 109-114 °C; Ref.: 15

R=Ac; R'=Bzl; 52%; m.p. 104-109 °C; [α]$_D$ -81.9 (chloroform); C-13; Ref.: 13, 16

R=R'=Ac; 51.2%; m.p. 109-115 °C; [α]$_D$ -84.8 (chloroform); C-13; Ref.: 6

II 50.9%; m.p. 111-114 °C; [α]$_D$ -59.9 (water); Ref.: 15

4-O-Me-α-D-GlcpA-(1→2)-β-D-Xylp-(1→4)-D-Xyl

β-D-(1-OMe) Me-ester; [α]$_D$ +19.5 (water); C-13; Ref.: 17

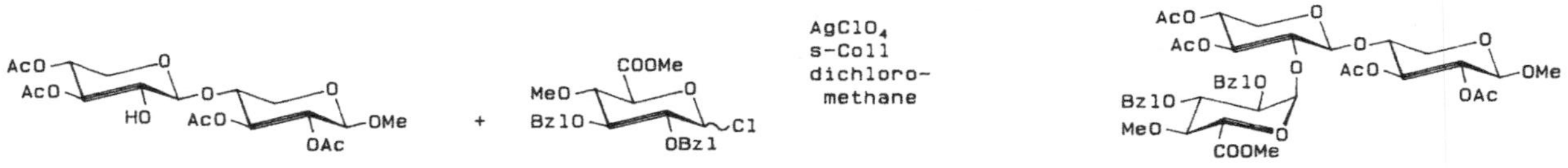

AgClO$_4$
s-Coll
dichloro-
methane

[α]$_D$ +5.2 (chloroform); Ref.: 17

4-O-Me-β-D-GlcpA-(1→2)-β-D-Xylp-(1→4)-D-Xyl

β-D-(1-OMe) Me-ester; m.p. 241-242°C; [α]$_D$ -66 (water); C-13; Ref.: 17

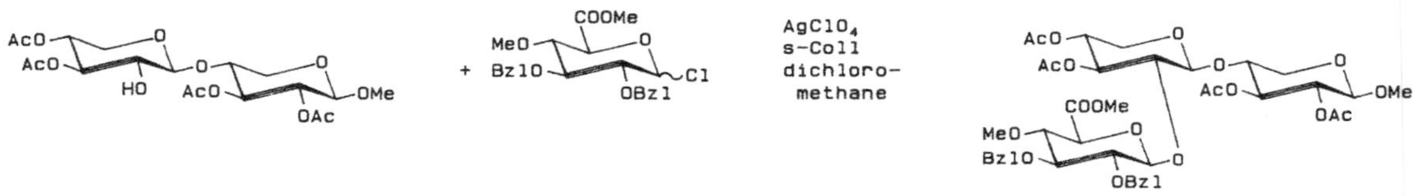

AgClO$_4$
s-Coll
dichloro-
methane

m.p. 163-164°C; [α]$_D$ -50 (chloroform); Ref.: 17

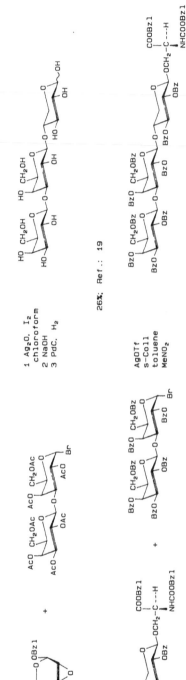

β-D-Galp-(1→3)-α-D-Galp-(1→4)-D-Xyl

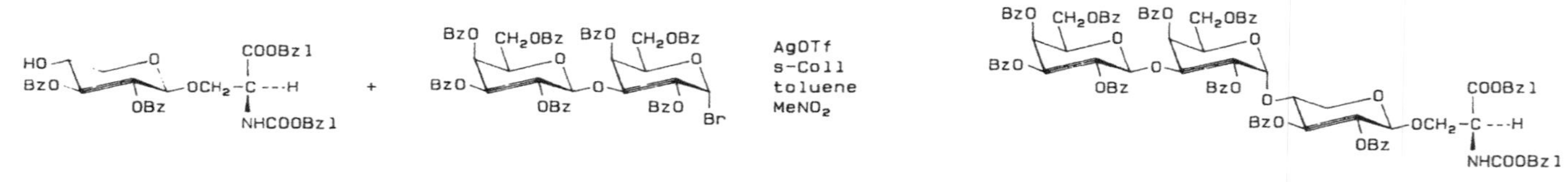

AgOTf
s-Coll
toluene
MeNO₂

10%; [α]_D +82 (chloroform); Ref.: 18

β-D-Galp-(1→3)-β-D-Galp-(1→4)-D-Xyl

amorphous; [α]_D +18 (water); Ref.: 19

β'-(1→3)-L-Serine; [α]_D -12 (water); C-13; Ref.: 18

1 Ag₂O, I₂
chloroform
2 NaOH
3 PdC, H₂

26%; Ref.: 19

AgOTf
s-Coll
toluene
MeNO₂

44%; [α]_D +37 (chloroform); C-13; Ref.: 18

α-L-Araf-(1→3)
β-D-Xylp-(1→4)   D-Xyl

β-D-(1-OMe): m.p. 206-209 °C; [α]_D -110 (water); C-13; Ref.: 10

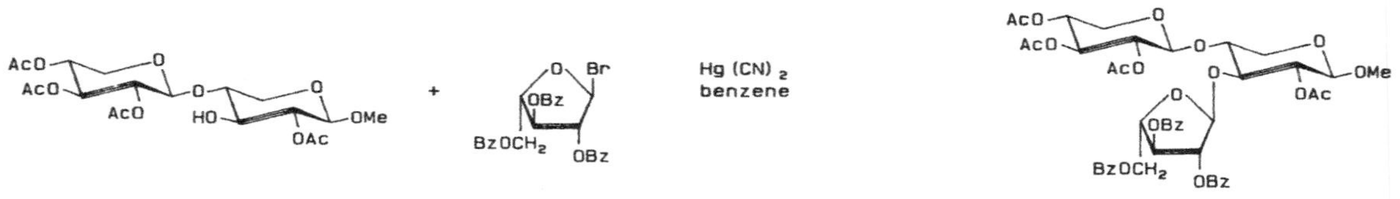

Hg(CN)_2
benzene

96.1%; m.p. 89-92 °C; [α]_D -44 (chloroform); C-13; Ref.: 10

α-D-Xylp-(1→2)
β-D-Xylp-(1→4)   D-Xyl

[α]_D +69.4 (water); C-13; Ref.: 20

Hg(CN)_2
MeCN

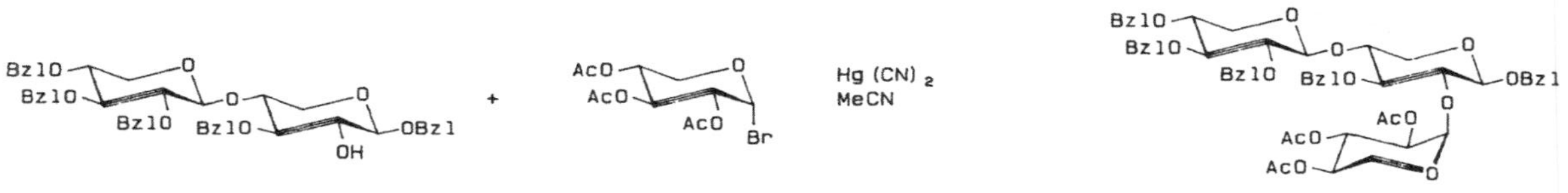

20%; m.p. 83-85 °C; [α]_D +3.2 (chloroform)

β-D-Xylp-(1-2)
β-D-Xylp-(1-4) ⟩D-Xyl

amorphous: [α]$_D$ +26 (water); C-13; Ref.: 20

β-D-(1-OMe): m.p. 185-193 °C; [α]$_D$ -79 (water); Ref.: 14

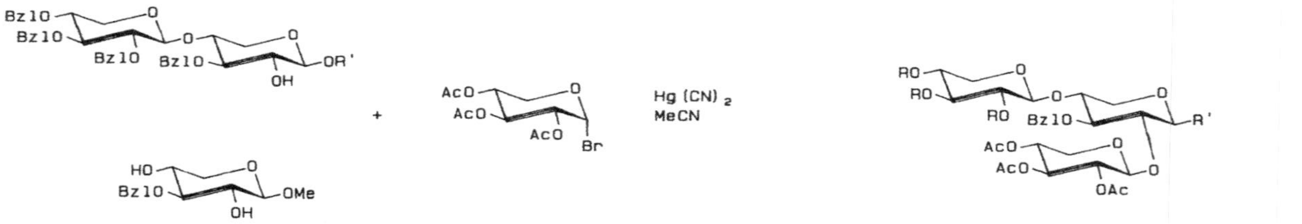

Hg(CN)$_2$
MeCN

R'=Me; R=Ac; 67.6%; m.p. 92-95 °C;
[α]$_D$ -82 (chloroform); Ref.: 14

R'=Me; R=Bzl: m.p. 126-128 °C;
[α]$_D$ -47 (chloroform); Ref.: 21

R'=R=Bzl: 70%; m.p. 126-127 °C;
[α]$_D$ -51.2 (chloroform); Ref.: 20

β-D-Xylp-(1→3)
β-D-Xylp-(1→4)　＼D-Xyl

β-D-(1→OMe): m.p. 216.5-217.5°C; [α]_D -76.2 (water): Ref.: 22

I Hg(CN)_2
   MeCN
II Ag_2CO_3, I_2
   chloroform

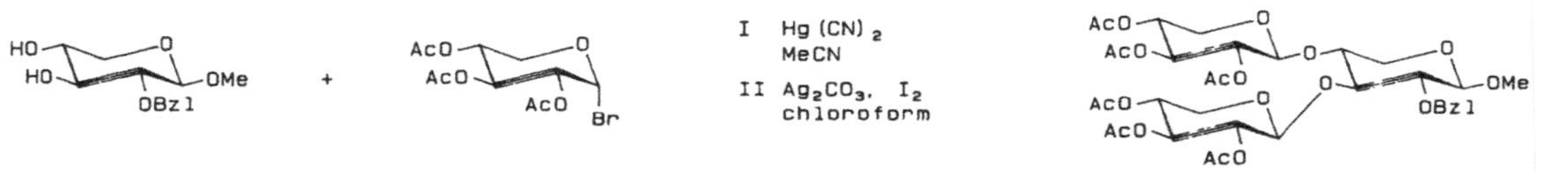

m.p. 164-166°C; [α]_D -73.8 (chloroform)

I 62%; Ref.: 22

II 39%; Ref.: 22

4-O-Me-α-D-GlcpA-(1→2)
β-D-Xylp-(1→4)　＼D-Xyl

β-D-(1→OMe) Me-ester; colorless, solid foam; [α]_D +32 (water); Ref.: 23

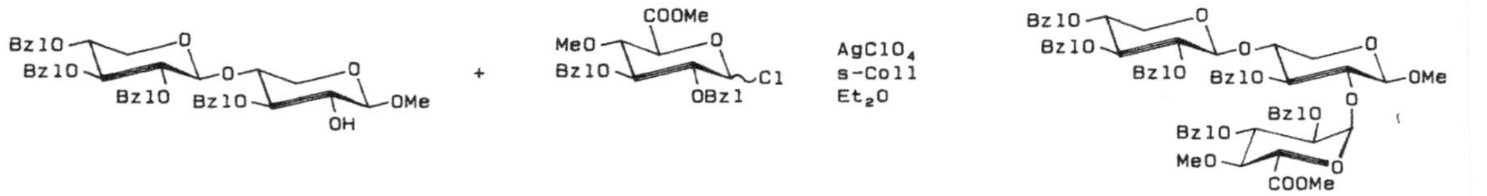

AgClO_4
s-Coll
Et_2O

87.1%; colorless, glassy solid; [α]_D +26.4 (chloroform)

4-O-Me-β-D-GlcpA-(1-2) ⟍
$\quad\quad\quad\quad\quad\quad\quad\quad\quad$ D-Xyl
β-D-Xylp-(1-4) ⟋

β-D-(1-OMe) Me-ester: m.p. 133-134 °C; [α]$_D$ -72.3 (water); Ref.: 23

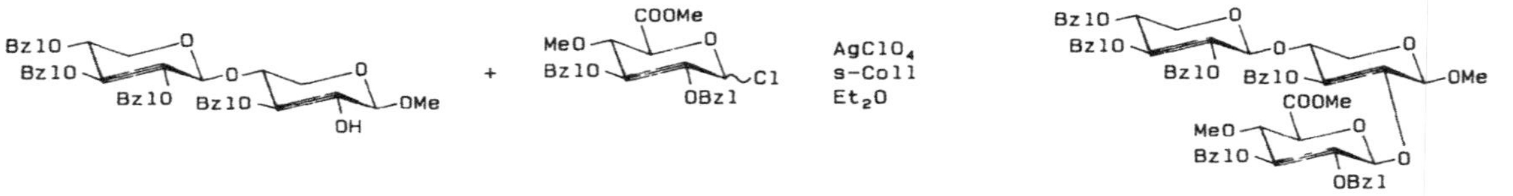

AgClO$_4$
s-Coll
Et$_2$O

11.3%: m.p. 96-99 °C; [α]$_D$ -16 (chloroform)

β-D-Glcp-(1-3) ⟍
$\quad\quad\quad\quad\quad\quad\quad$ D-Xyl
β-D-Glcp-(1-5) ⟋

hygroscopic glass; [α]$_D$ -30 (water); Ref.: 24

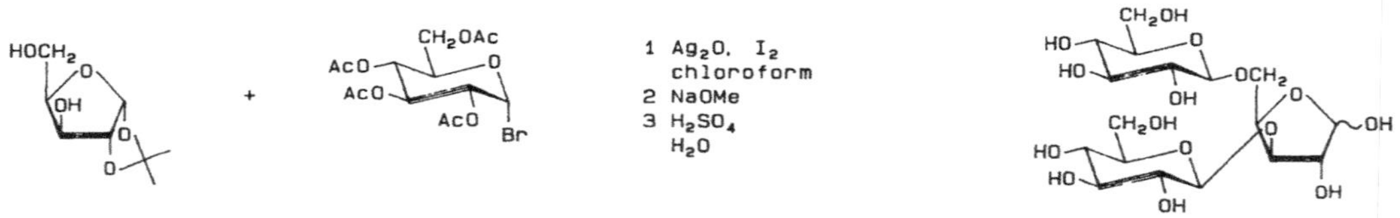

1 Ag$_2$O, I$_2$
   chloroform
2 NaOMe
3 H$_2$SO$_4$
   H$_2$O

α-D-Glcp-(1→4)-α-D-Glcp-(1→1)-α-D-Glcp

[α]$_D$ +169 (water): Ref.: 25

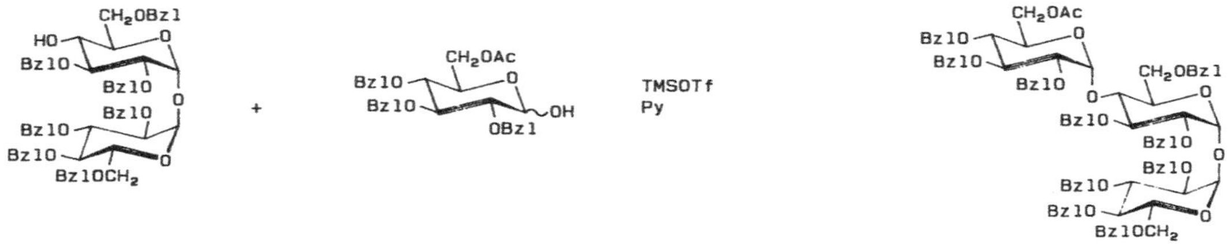

+

TMSOTf
Py

76%: [α]$_D$ +81 (chloroform)

α-D-Glcp-(1→6)-α-D-Glcp-(1→1)-α-D-Glcp

[α]$_D$ +144 (water): Ref.: 25

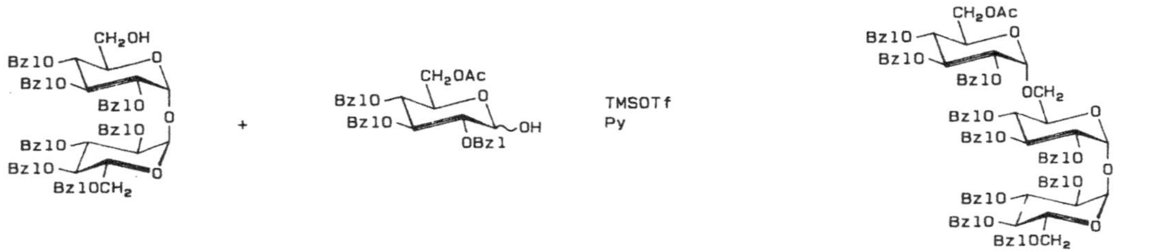

+

TMSOTf
Py

79%: [α]$_D$ +93 (chloroform)

α-D-Glcp-(1→4)-α-D-Glcp-(1→1)-α-D-Glcp

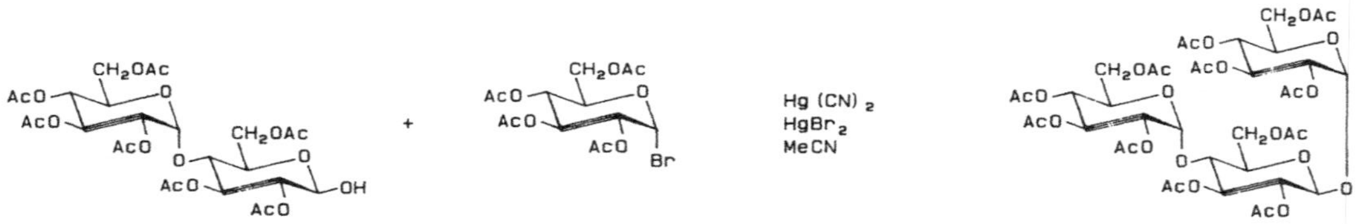

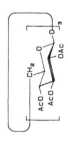

+

Hg(CN)₂
HgBr₂
MeCN

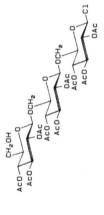

18.4%; m.p. 195—197 °C; [α]$_D$ +101.8 (chloroform); Ref.: 26

→6)-[β-D-Glcp]₃-(1→

HgBr₂
dichloro-
ethane

16%; m.p. 285 °C; [α]$_D$ −11 (chloroform); Ref.: 27

α-D-Glcp-(1→2)-α-D-Glcp-(1→2)-D-Glc

monohydrate; m.p. 228-230°C (dec.); [α]$_D$ +150.2 → +156.1 (water); Ref.: 28

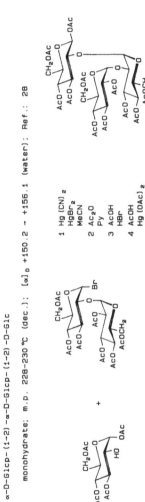

1 Hg(CN)$_2$
  HgBr$_2$
  MeCN
2 Ac$_2$O
  Py
3 AcOH
  HBr
4 AcOH
  Hg(OAc)$_2$

23%; m.p. 187-188°C; [α]$_D$ +123.1 (chloroform)

α-L-Araf-(1→6)-α-D-Glcp-(1→4)-D-Glc

β-D-(1-OMe); white powder; [α]$_D$ +4 (water); C-13; Ref.: 29

Hg(CN)$_2$
MeNO$_2$

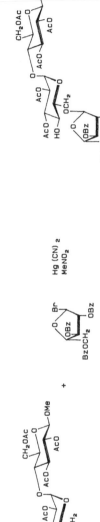

62%; m.p. 85-86°C; [α]$_D$ +20.1 (chloroform)

α-D-Glcp-(1→6)-α-D-Glcp-(1→4)-D-Glc

m.p. 221°C; Ref.: 30

m.p. 221°C; [α]_D +161 → +151 (water); Ref.: 31

m.p. 220-222°C; [α]_D +161.1 → 151 (water); Ref.: 32

m.p. 195-198°C (dec.); [α]_D +147 (water); C-13; Ref.: 33

m.p. 219-220C (dec.): [α]_D +162.1 → 151.4 (water); Ref.: 34

I   1 AgClO₄ MeNO₂
    2 H₂; PdC
    3 NaOMe

II 1 AgClO₄ Ag₂CO₃ Et₂O
    2 H₂; PdC
    3 NaOMe

III 1 Hg-succinate benzene
    2 Ac₂O; Py
    3 NaOMe
    4 H₂; PdC

IV Et₄NBr
    DMF
    dichloromethane

V   CoBr₂
    (nBu)₄NBr
    Me₃SiBr
    dichloromethane

I   R=OAc; R¹=H; R²=Ac; R³=Tr; Ref.: 30
    R=H; R¹=OAc; R²=Ac; R³=Tr; 17%; Ref.: 31

II R=H; R¹=OAc; R²=Ac; R³=H; 6%; Ref.: 31

III R=H; R¹=OBzl; R²=Ac; R³=H; 40.5%; Ref.: 32

IV R=Ac; R¹=H; R²=OAc; R³=H; 71%; m.p. 152-153°C; [α]_D +51.3 (chloroform); C-13; Ref.: 34

V   R=R³=Bzl; R¹=OBzl; R²=H; 29%; [α]_D +75 (chloroform); C-13; Ref.: 33

β-D-Glcp-(1→6)-α-D-Glcp-(1→4)-D-Glc

[α]_D +70 (water); Ref.: 31

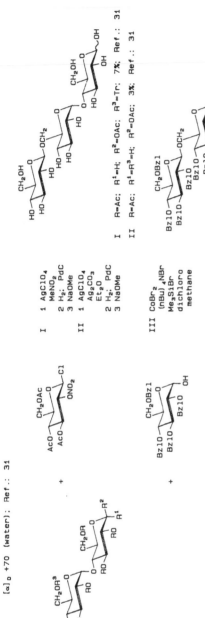

I   1 AgClO_4
      MeNO_2
    2 H_2: PdC
    3 NaOMe

II 1 AgClO_4
      Ag_2CO_3
      Et_2O
    2 H_2: PdC
    3 NaOMe

III CoBr_2
    (nBu)_4NBr
    Me_3SiBr
    dichloro
    methane

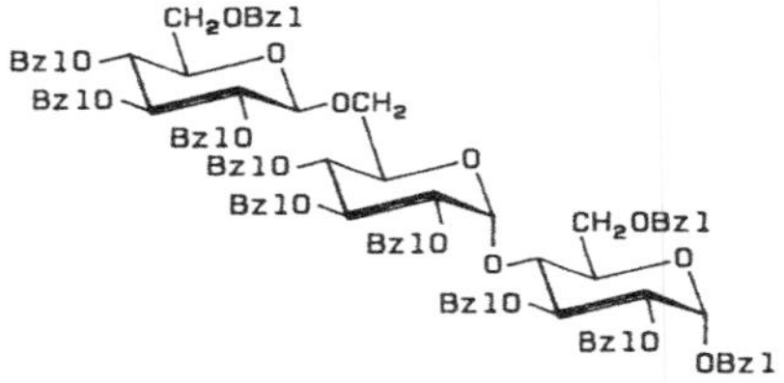

I   R=AC; R^1=H; R^2=OAc; R^3=Tr; 7%; Ref.: 31

II   R=AC; R^1=R^3=H; R^2=OAc; 3%; Ref.: 31

III R=Bzl; R^2=R^3=H; R^1=OBzl; 9%;
     [α]_D +57 (chloroform); C-13; Ref.: 33

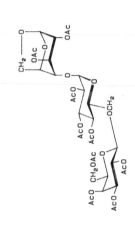

α-D-Galp-(1→6)-α-D-Glcp-(1→4)-D-Glc

amorphous powder: [α]_D +159.2 (water): Ref.: 35

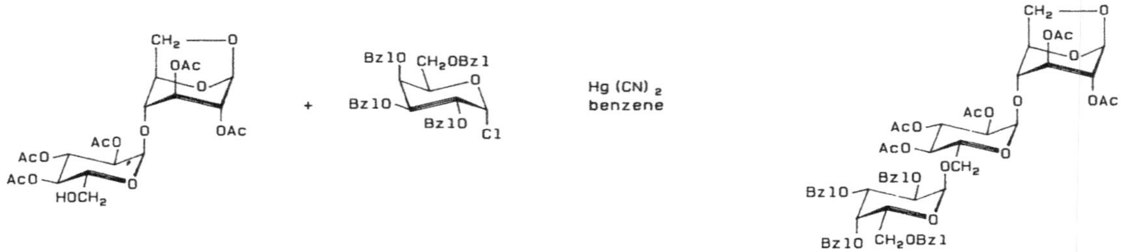

Hg(CN)₂
benzene

63%; m.p. 59-60 °C; [α]_D +48.2 (chloroform)

α-D-Galp-(1→6)-α-D-Glcp-(1→4)-D-Glc

amorphous powder: [α]_D +85 (water): Ref.: 35

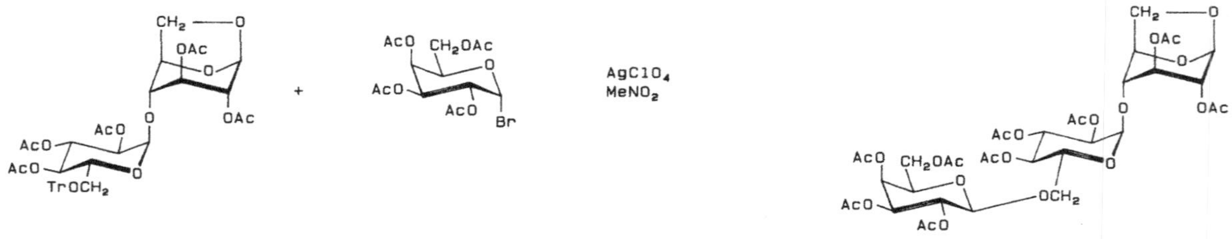

AgClO₄
MeNO₂

56%; m.p. 217-218 °C; [α]_D +19.7 (chloroform)

4-Amino-4,6-Dideoxy-α-D-Glcp-(1→4)-α-D-Glcp-(1→4)-D-Glc

α-p-(1→OPh) N-2,3,4-trimethoxybenzoyl: $[\alpha]_{578}$ +161 (MeOH): Ref.: 36

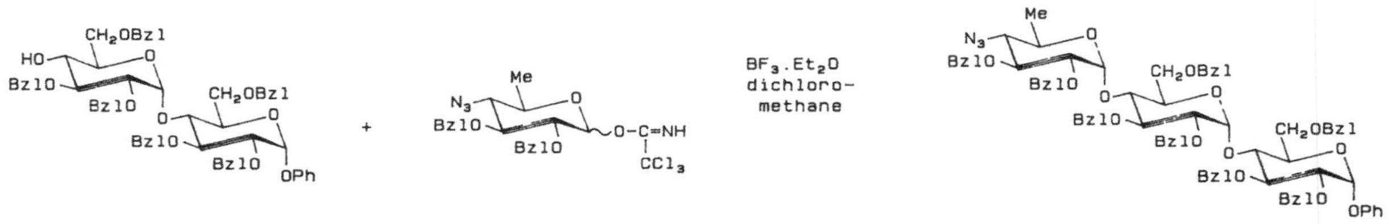

BF$_3$.Et$_2$O
dichloro-
methane

22%: $[\alpha]_{578}$ +106 (chloroform)

4-Amino-4,6-Dideoxy-β-D-Glcp-(1→4)-α-D-Glcp-(1→4)-D-Glc

α-p-(1→OPh) N-2,3,4-trimethoxybenzoyl: $[\alpha]_{578}$ +121 (MeOH): Ref.: 36

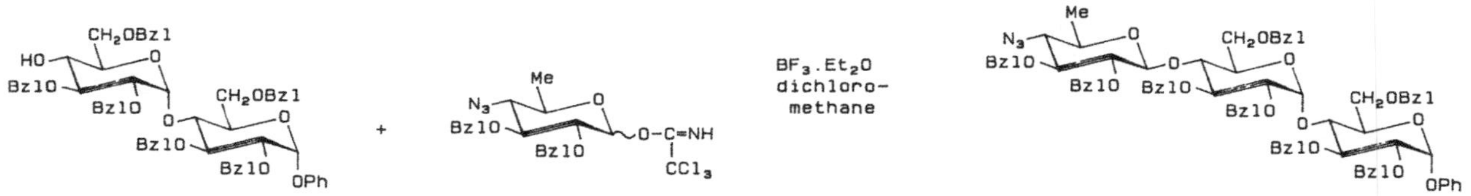

BF$_3$.Et$_2$O
dichloro-
methane

22%: $[\alpha]_{578}$ +72 (chloroform)

α-D-Glcp-(1→2)-α-D-Glcp-(1→6)-D-Glc

white, amorphous powder: $[\alpha]_D$ +150.5 (water); C-13; Ref.: 37

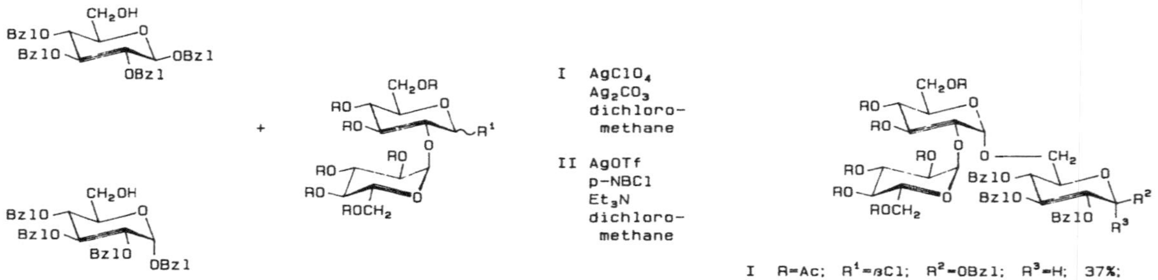

I AgClO$_4$
   Ag$_2$CO$_3$
   dichloro-
   methane

II AgOTf
   p-NBCl
   Et$_3$N
   dichloro-
   methane

I R=Ac; R^1=βCl; R^2=OBzl; R^3=H; 37%;
   $[\alpha]_D$ +86.5 (chloroform); Ref.: 37

II R=Bzl; R^1=OH; R^2=H; R^3=OBzl; 51%;
   $[\alpha]_D$ +86 (chloroform); Ref.: 38

α-D-Glcp-(1→3)-α-D-Glcp-(1→6)-D-Glc

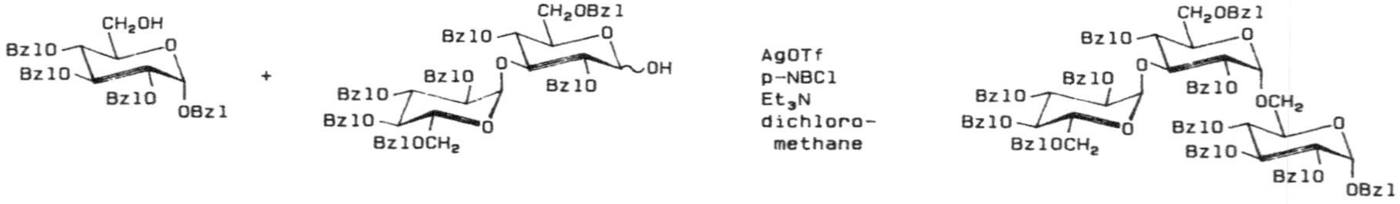

AgOTf
p-NBCl
Et$_3$N
dichloro-
methane

67%; $[\alpha]_D$ +80 (chloroform); Ref.: 38

α-D-Glcp-(1→4)-α-D-Glcp-(1→6)-D-Glc

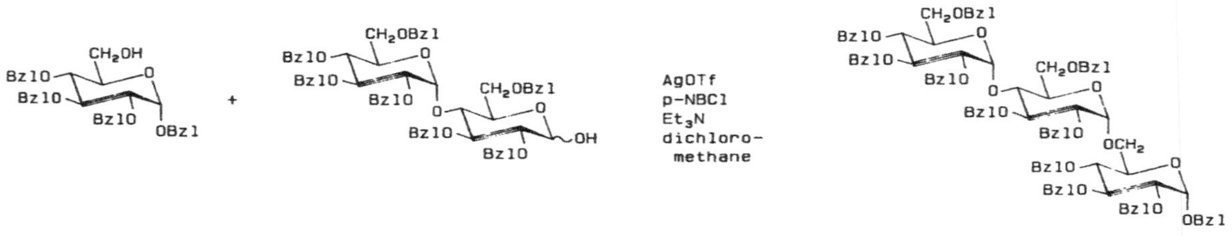

AgOTf
p-NBCl
Et₃N
dichloro-
methane

58%: [α]_D +74 (chloroform); Ref.: 38

α-D-Glcp-(1→6)-α-D-Glcp-(1→6)-D-Glc

amorphous powder: $[\alpha]_D$ +139 (water): Ref.: 39

I   Et₂O

II   Hg(CN)₂
     MeNO₂

III Hg(CN)
     HgBr₂
     dichloro-
     ethane

IV   AgOTf
     p-NBCl
     Et₃N
     dichloro-
     methane

I   R=R³=R⁴=Bzl; R¹=OMe; R²=H; R⁵=CO-NHPh; 85%; Ref.: 41
     R=R³=R⁴=Bzl; R¹=OMe; R²=H; R⁵=CO-NHPh; 85%;
     [α]_D +74.5 (chloroform): Ref.: 40
     R=R³=R⁴=Bzl; R¹=O(CH₂)₂-p-Ph-NHTs; R²=H; R⁵=CO-NHPh; 85%;
     [α]_D +64.1 (chloroform): Ref.: 42

II   R=Ac; R¹=H; R²=OAc; R³=Bzl; R⁴=R⁵=NBz; R⁶=Br; 61%;
     [α]_D +82.4 (chloroform): Ref.: 39

III R=R³=R⁴=Bzl; R¹=H; R²=OAll; R⁵=MCA; R⁶=Cl;
     83%; C-13; Ref.: 43

IV   R=R³=R⁴=R⁵=Bzl; R¹=OBzl; R²=H; 72%;
     [α]_D +81 (chloroform); Ref.: 38

β-D-Glcp-(1→2)-α-D-Glcp-(1→6)-D-Glc

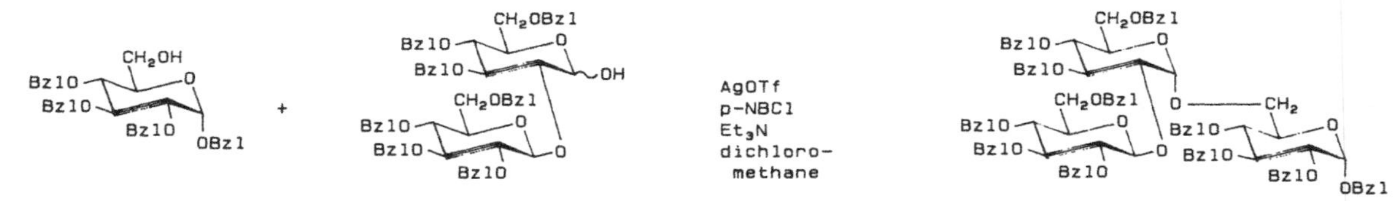

44%; [α]$_D$ +59 (chloroform); Ref.: 38

β-D-Glcp-(1→3)-α-D-Glcp-(1→6)-D-Glc

m.p. 149—150 °C; [α]$_D$ +48 → +57 (water); C-13; Ref.: 44

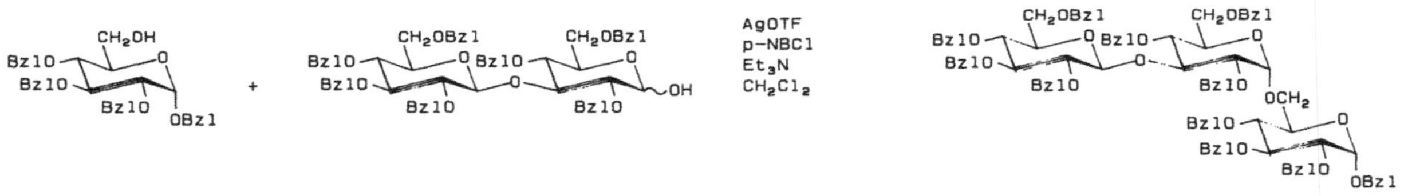

38%; [α]$_D$ +67 (chloroform); Ref.: 38, 44

β-D-Glcp-(1→4)-α-D-Glcp-(1→6)-D-Glc

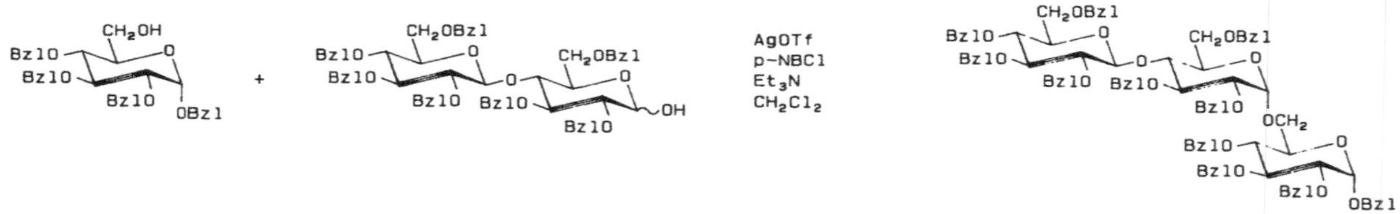

AgOTf
p-NBCl
Et₃N
CH₂Cl₂

16%; [α]_D +62 (chloroform); Ref.: 38, 44

β-D-Glcp-(1→6)-α-D-Glcp-(1→6)-D-Glc

[α]_D +71.9 (water); Ref.: 39

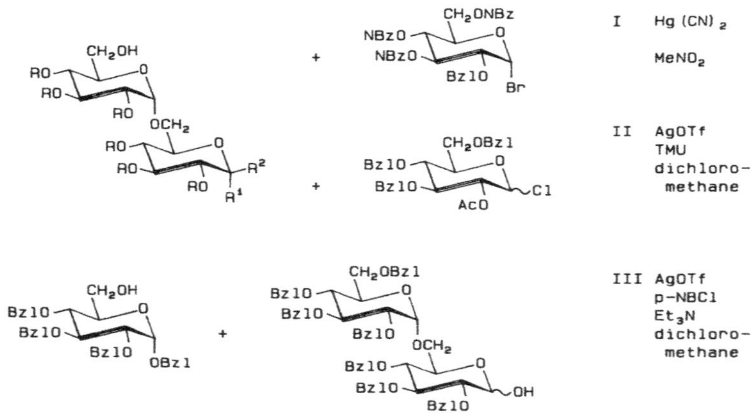

I   Hg(CN)₂
     MeNO₂

II   AgOTf
     TMU
     dichloro-
     methane

III AgOTf
     p-NBCl
     Et₃N
     dichloro-
     methane

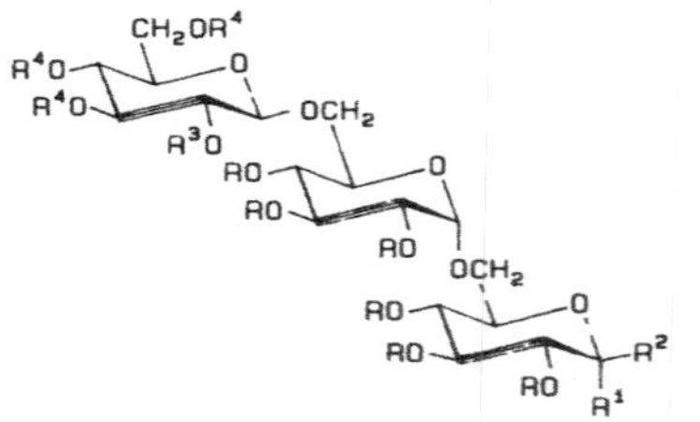

I   R=Ac; R¹=H; R²=OAc; R³=Bzl; R⁴=NBz; 17%;
     [α]_D +46.5 (chloroform); Ref.: 39

II   R=R⁴=Bzl; R¹=N₃; R²=H; R³=Ac; 48.8%;
     [α]_D +64.9 (chloroform); C-13; Ref.: 45

III R=R³=R⁴=Bzl; R¹=OBzl; R²=H; 63%;
     [α]_D +53 (chloroform); Ref.: 38

β-D-Galp-(1-4)-α-D-Glcp-(1-6)-D-Glc

α-D-Glcp-(1-2)-β-D-Glcp-(1-2)-D-Glc

monohydrate: m.p. 182-183°C (dec.); [α]$_D$ +78.5 → +72.8 (water); Ref.: 28

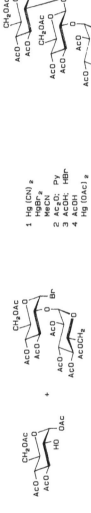

20%; [α]$_D$ +53 (chloroform); Ref: 44

12%; m.p. 179-180 °C; [α]$_D$ +78.5 (chloroform); Ref.: 28

β-D-Glcp-(1→2)-β-D-Glcp-(1→2)-D-Glc

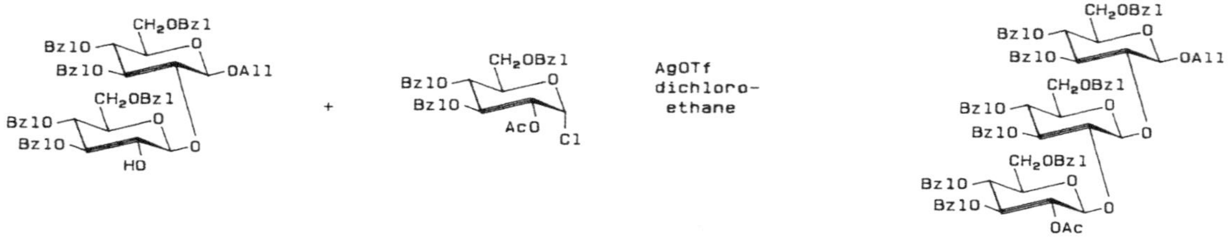

AgOTf
dichloro-
ethane

98%; $[\alpha]_D$ -0.56 (chloroform); C-13; Ref.: 46

β-D-Glcp-(1→4)-β-D-Glcp-(1→3)-D-Glc

m.p. 234—237 °C; $[\alpha]_D$ +17 → +12 (water); Ref.: 47

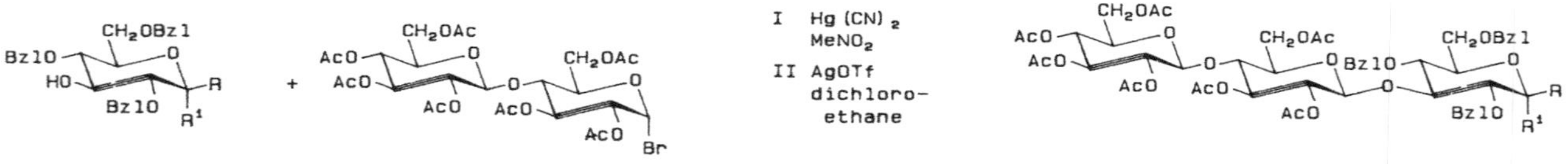

I Hg(CN)₂
  MeNO₂
II AgOTf
  dichloro-
  ethane

I R=H; R¹=OMe; 71%; m.p. 65—66 °C;
   $[\alpha]_D$ +16 (chloroform); C-13; Ref.: 48

II R=OBzl; R¹=H; 82%; $[\alpha]_D$ -23 (chloroform); C-13; Ref.: 47

β-D-Glcp-(1→3)-β-D-Glcp-(1→4)-D-Glc

m.p. 227—230 °C; [α]$_D$ +18 → +13 (water); Ref.: 47

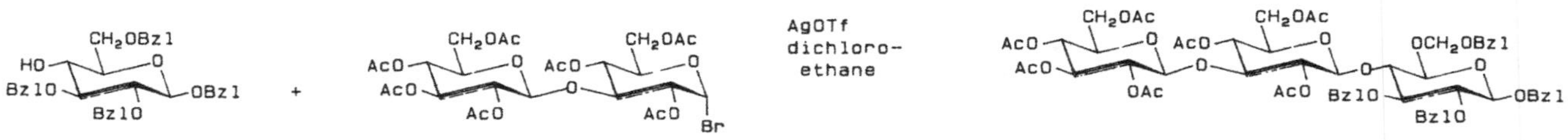

BO%; [α]$_D$ −25 (chloroform); C-13; Ref.: 47

α-D-Glcp-(1→4)-β-D-Glcp-(1→4)-D-Glc

m.p. 206-208.5°C (dec.) $[\alpha]_D$ +32.8 → +21.0 (water); Ref.: 50

α-p-(1-OR:

R=   ⎯OH Me $[\alpha]_D$ +58.0 (water); C-13; Ref.: 53, 54

OH ⎯Me $[\alpha]_D$ +65.0 (water); C-13; Ref.: 53, 54

H OH Me $[\alpha]_D$ +44.3 (water); C-13; Ref.: 53, 54

OH H Me $[\alpha]_D$ +62.0 (water); C-13; Ref.: 53, 54

I   +   (AcO ··· CH$_2$OAc ··· Br)   →   1 AgOTf dichloro-methane   2 NaOMe

R^2=H:

R^1=O ⎯OBzl Me   55-68%; $[\alpha]_D$ +39.1 (chloroform); C-13; Ref.: 53, 54

R^1=O OBzl ⎯Me   63-82%; $[\alpha]_D$ +25.0 (chloroform); C-13; Ref.: 53, 54

R^1=O H ⎯OBzl Me 31-63.5%; $[\alpha]_D$ +50.0 (chloroform); C-13; Ref.: 53, 54

R^1=O ⎯OBzl H Me 30-67.9%; $[\alpha]_D$ +37.5 (chloroform); C-13; Ref.: 53, 54

I

II   Ag$_2$O: chloroform

III   Ag$_2$CO$_3$: chloroform

II   R=Me; m.p. 210—211 °C; [α]$_D$ +34.2 (chloroform); Ref.: 51

III   R=Me; m.p. 210—211 °C; [α]$_D$ +34.5 (chloroform); Ref.: 51

IV   Hg(CN)$_2$ dichloro-ethane

IV   40%; [α]$_D$ −29.3 (chloroform); Ref.: 52

V   AgOTf TMU dichloro-methane

V   R=Ac; m.p. 209—210 °C; [α]$_D$ −18.0 (chloroform); Ref.: 52

VI   1 Hg(CN)$_2$ benzene MeNO$_2$ 2 NaOMe

VI   R^1=H; R^2=OBzl; 77% m.p. 193—194 °C; [α]$_D$ +4.9 (chloroform); Ref.: 50

R^1=H; R^2=OMe; 80%; [α]$_D$ +15.7 (chloroform); Ref.: 50

R^1=OMe; R^2=H; 77%; [α]$_D$ +24.7 (chloroform); Ref.: 50

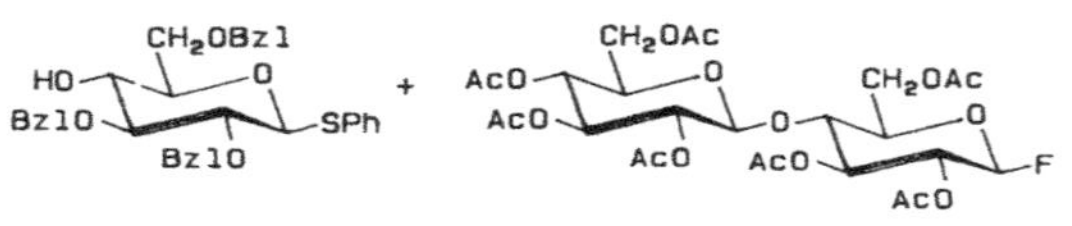

VII 75%: Ref.: 49

VIII 57%: Ref.: 55

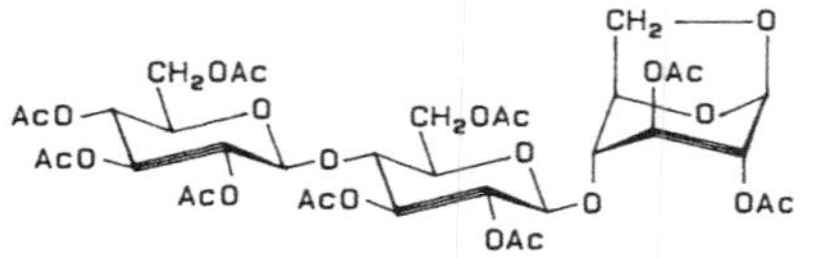

V 53%: [α]$_D$ −29.4 (chloroform): Ref.: 52

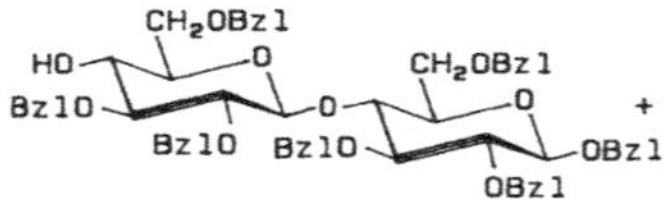

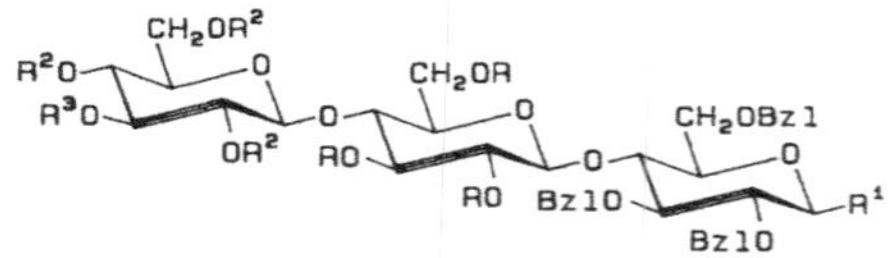

VI R=Bzl; R^1=OBzl; R^2=H; R^3=H; 80%:
[α]$_D$ +9.3 (chloroform): Ref.: 50

IX R=Bzl; R^1=OBzl; R^2=Ac; R^3=All; 83%:
[α]$_D$ −15 (chloroform): Ref.: 47

VII AgClO$_4$
SnCl$_2$
dichloro-
methane

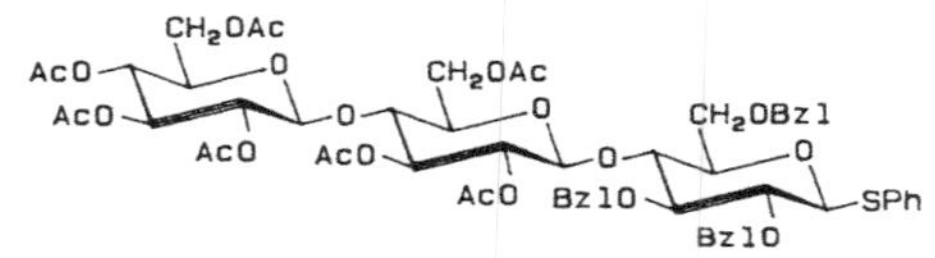

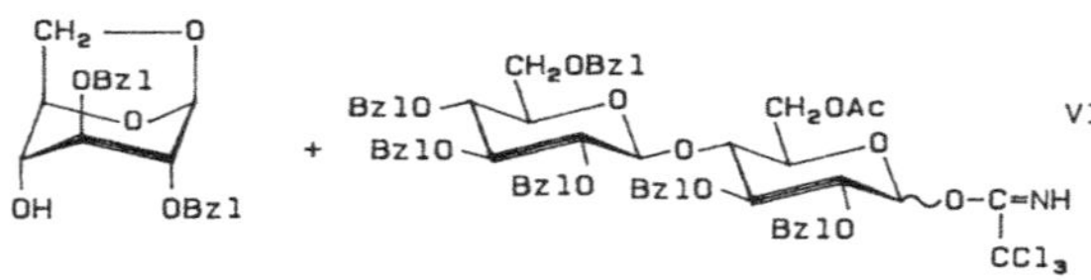

VIII BF$_3$.Et$_2$O
dichloro-
methane

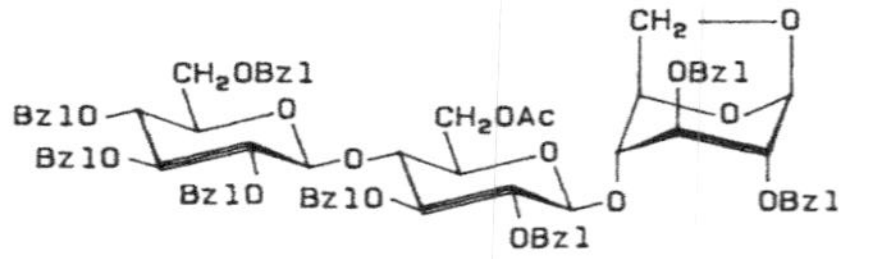

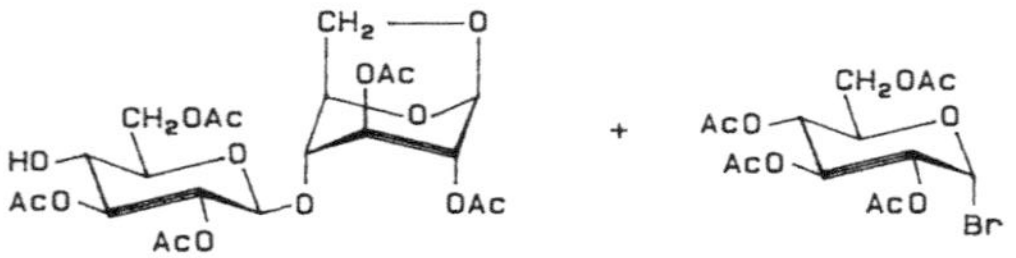

V AgOTf
TMU
dichloro-
methane

VI 1 Hg(CN)$_2$
benzene
MeNO$_2$
2 NaOMe

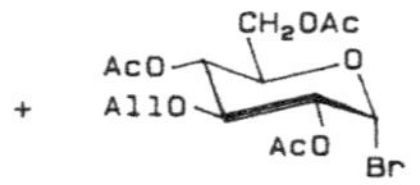

IX AgOTf
dichloro-
ethane

β-D-GlcpA-(1→6)-β-D-Glcp-(1→4)-D-Glc

[α]_D -4.8 (water); Ref.: 56

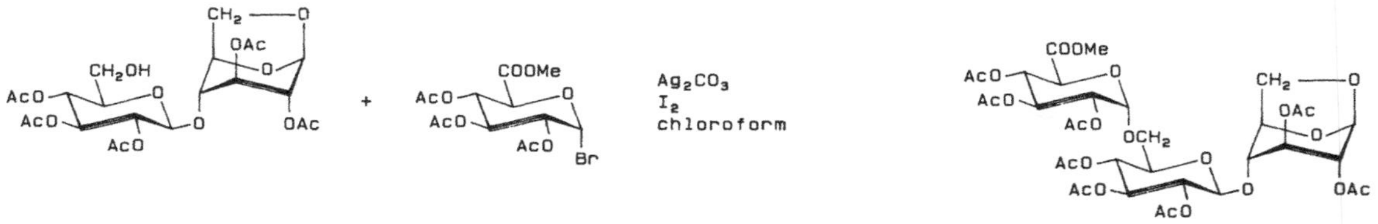

85%; m.p. 210—211°C; [α]_D -54.2 (chloroform); Ref.: 56

β-D-Galp-(1→4)-β-D-Glcp-(1→4)-D-Glc

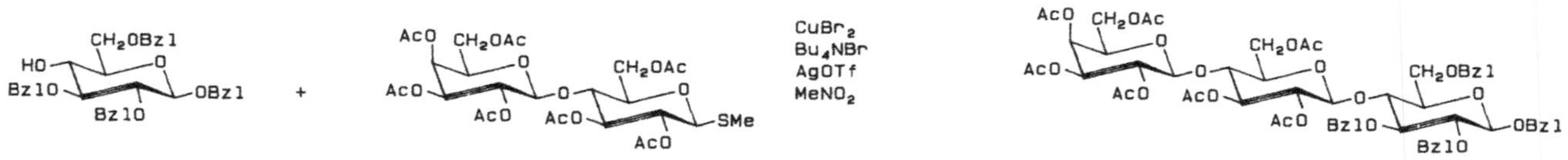

71%; [α]_D -12 (chloroform); Ref.: 57

α-D-Glcp-(1→2)-β-D-Glcp-(1→6)-D-Glc

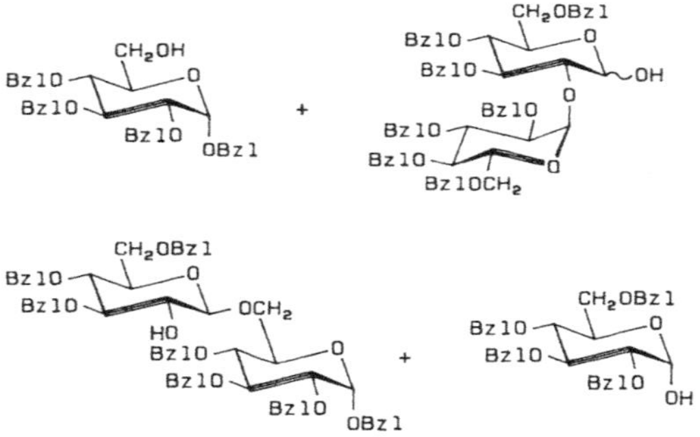

Ref.: 38

AgOTf
p-NBCl
Et₃N
dichloro-
methane

Ref.: 58

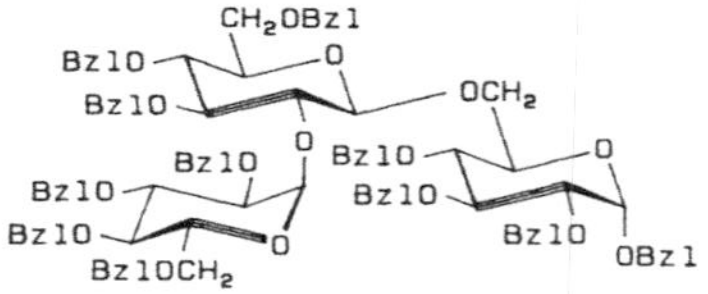

51%; $[\alpha]_D$ +67 (chloroform); Ref.: 38

21%; $[\alpha]_D$ +67 (chloroform); C-13; Ref.: 58

α-D-Glcp-(1→3)-β-D-Glcp-(1→6)-D-Glc

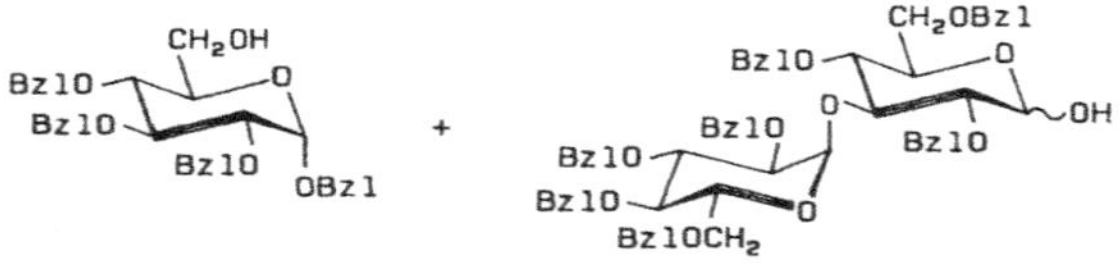

AgOTf
p-NBCl
Et₃N
dichloro-
methane

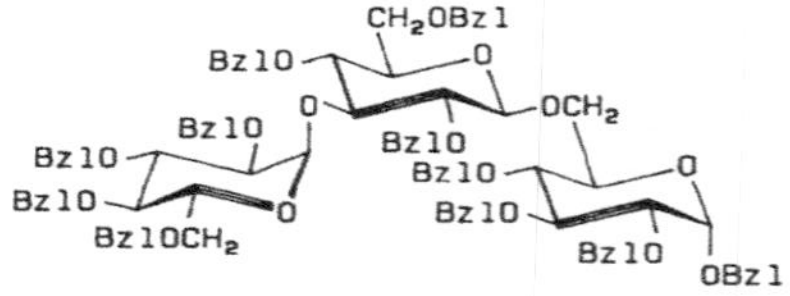

67%; $[\alpha]_D$ +56 (chloroform); Ref.: 38

α-D-Glcp-(1→4)-α-D-Glcp-(1→6)-D-Glc

amorphous; $[\alpha]_D$ +62 (water); Ref.: 62

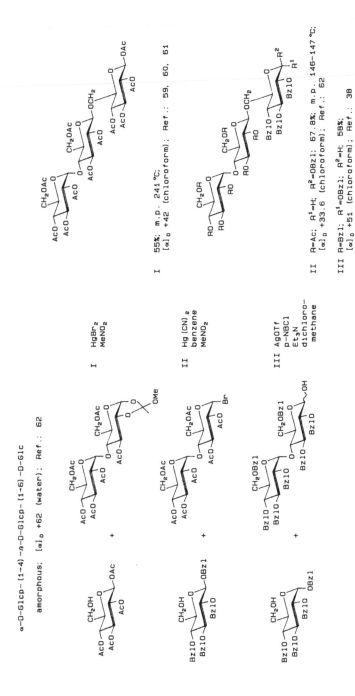

I   HgBr₂
     MeNO₂

II   Hg(CN)₂
     benzene
     MeNO₂

III AgOTf
     p-NBCl
     Et₃N
     dichloro-
     methane

I   55%; m.p. 241 °C;
     $[\alpha]_D$ +42 (chloroform); Ref.: 59, 60, 61

II   R=Ac; R¹=H; R²=OBzl; 67.8%; m.p. 146—147 °C;
     $[\alpha]_D$ +33.6 (chloroform); Ref.: 62

III R=Bzl; R¹=OBzl; R²=H; 58%;
     $[\alpha]_D$ +51 (chloroform); Ref.: 38

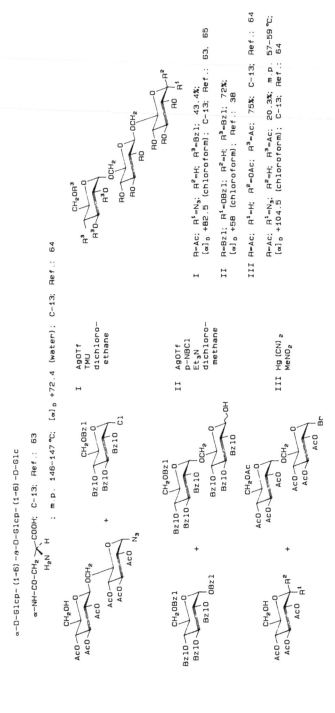

α-D-Glcp-(1-6)-β-D-Glcp-(1-6)-D-Glc

α-NH-CO-CH₂—COOH; C-13; Ref.: 63
H₂N H   : m.p. 146-147 °C; [α]_D +72.4 (water); C-13; Ref.: 64

I   AgOTf
    TMU
    dichloro-
    ethane

II AgOTf
    p-NBCl
    Et₃N
    dichloro-
    methane

III Hg(CN)₂
    MeNO₂

I   R=Ac; R¹=N₃; R²=H; R³=Bzl; 43.4%;
     [α]_D +82.5 (chloroform); C-13; Ref.: 63, 65

II   R=Bzl; R¹=OBzl; R²=H; R³=Bzl; 72%;
     [α]_D +58 (chloroform); Ref.: 38

III R=Ac; R¹=H; R²=OAc; R³=Ac; 75%; C-13; Ref.: 64

     R=Ac; R¹=N₃; R²=H; R³=Ac; 20.3%; m.p. 57-59 °C;
     [α]_D +104.5 (chloroform); C-13; Ref.: 64

α-D-Glcp-(1-2)-β-D-Glcp-(1-6)-D-Glc

glassy; [α]$_D$ -1 (water); Ref.: 58

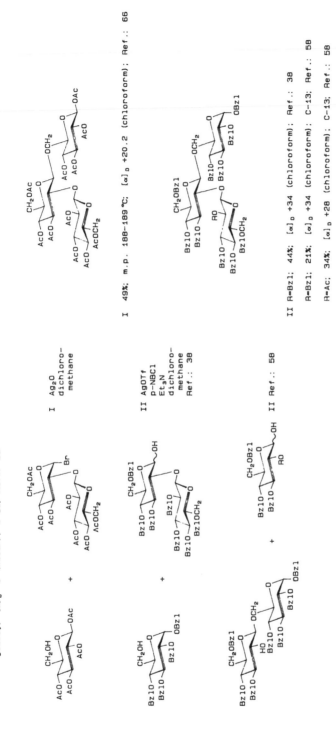

I   Ag$_2$O
    dichloro-
    methane

II AgOTf
    p-NBCl
    Et$_3$N
    dichloro-
    methane
    Ref.: 38

II Ref.: 58

I   49%; m.p. 188-189 °C; [α]$_D$ +20.2 (chloroform); Ref.: 66

II R=Bzl; 44%; [α]$_D$ +34 (chloroform); Ref.: 38
    R=Bzl; 21%; [α]$_D$ +34 (chloroform); C-13; Ref.: 58
    R=Ac; 34%; [α]$_D$ +28 (chloroform); C-13; Ref.: 58

β-D-Glcp-(1→3)-β-D-Glcp-(1→6)-D-Glc

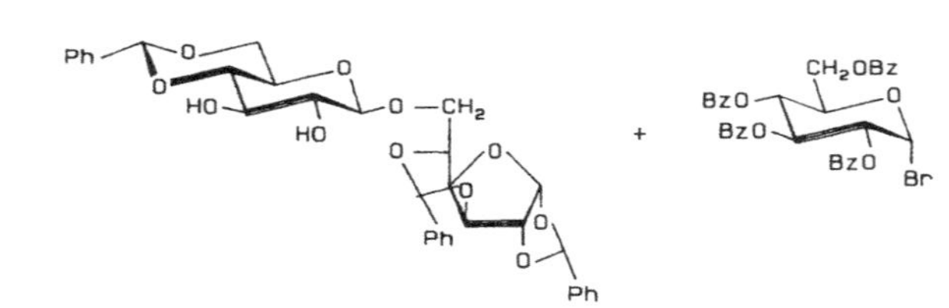

I   1 AgOTf
    s-Coll
    toluene
    2 Ac₂O: Py

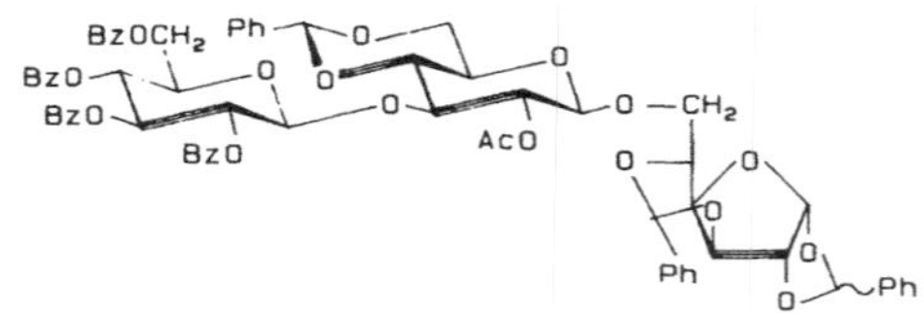

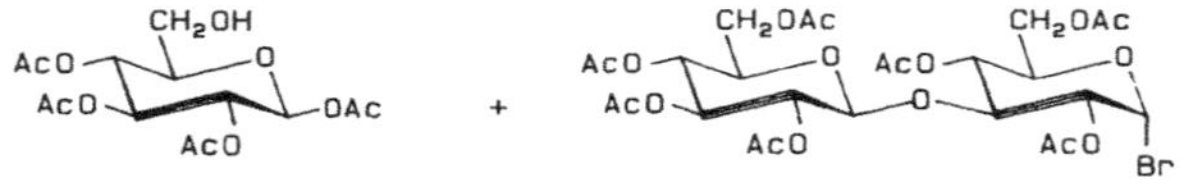

II Ag₂O
    dichloro-
    methane

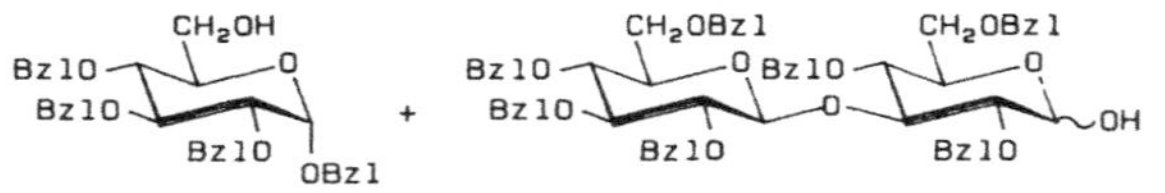

III AgOTf
    p-NBCl
    Et₃N
    dichloro-
    methane

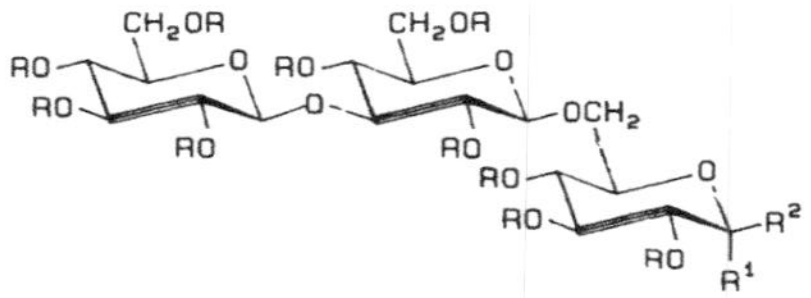

I   90% (57% exo, 33% endo):
    endo $[\alpha]_D$ -0.3 (dichloromethane): Ref.: 67, 68

II R=Ac: R^1=H; R^2=OAc: 74%; m.p. 216-217 °C;
    $[\alpha]_D$ -27.5 (chloroform): Ref.: 66

III R=Bzl; R^1=OBzl; R^2=H; 24%;
    $[\alpha]_D$ +42 (chloroform): Ref.: 38, 44

α-D-Glcp-(1-4)-α-D-Glcp-(1-6)-D-Glc

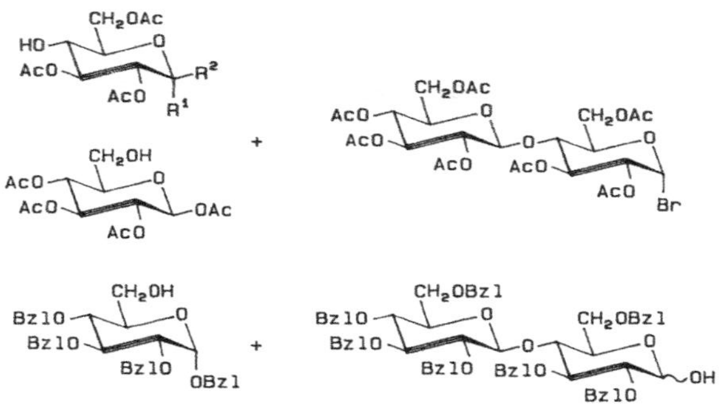

I   Ag₂O; I₂
chloroform

II Ag₂O
dichloro-
methane

III AgOTf
p-NBCl
Et₃N
dichloro-
methane

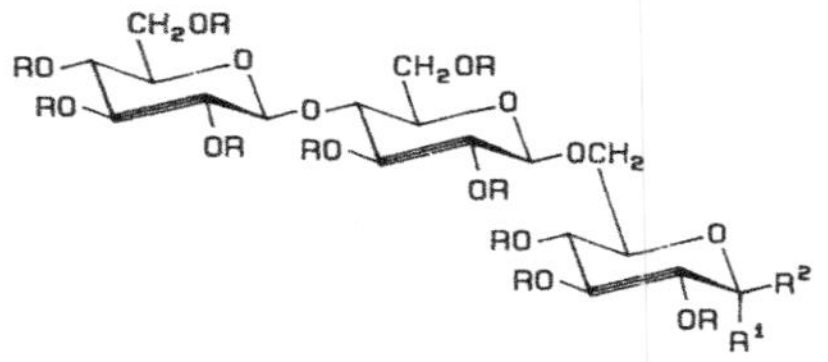

I   R=Ac; R¹=OAc; R²=H; m.p. 190—191 °C;
[α]_D −17.5 (chloroform): Ref.: 69

R=Ac; R¹=H; R²=OAc; m.p. 195—195.5 °C;
[α]_D +30.9 (chloroform): Ref.: 69

II R=Ac; R¹=H; R²=OAc; 70%; m.p. 240—241 °C;
[α]_D −9.7 (chloroform): Ref.: 66

III R=Bzl; R¹=OBzl; R²=H; 49%; m.p. 122—123 °C;
[α]_D +48 (chloroform): Ref.: 38, 44

β-D-Glcp-(1-6)-β-D-Glcp-(1-6)-D-Glc

amorphous powder: m.p. 143-151 °C; $[α]_D$ -1.1 (water): Ref.: 70

β-p-(1-0-Ph-OH): m.p. 205-207 °C; $[α]_D$ -59.5 (water): Ref.: 71

I Ag₂CO₃: I₂

II Hg(CN)₂: HgBr₂

III Ag₂O dichloromethane

IV Ag₂O: I₂ chloroform

V 1 AgClO₄: AllBr benzene 2 Ac₂O: NaOAc

I R¹=H: R²=OAc

VI AgOTf TMU dichloromethane

VII AgOTf p-NBCl Et₃N dichloromethane

I R=R¹=Ac: 52% (1+2): Ref.: 70
R=R¹=Ac: 43% (2+1): m.p. 221-223 °C; $[α]_D$ -7.4 (chloroform): Ref.: 70

II R=R¹=Ac: 68%: m.p. 220-221 °C; $[α]_D$ -7 (chloroform): Ref.: 72
R=TCA: R¹=Ac: 64%: m.p. 200 °C; $[α]_D$ -3 (chloroform): Ref.: 72

III R=R¹=Ac: 39%: m.p. 220-221 °C; $[α]_D$ -8 (chloroform): Ref.: 66

IV R=Ac: R¹=PhOAc: 59%: m.p. 241-242 °C; $[α]_D$ -24.3 (chloroform): Ref.: 71

V R=R¹=Ac: 12%: m.p. 215-216 °C; $[α]_D$ -7 (chloroform): Ref.: 73

VI R=Ac: R¹=N₃: 30.3%; $[α]_D$ +61.3 (chloroform): C-13: Ref.: 63, 65

VII R=Bzl: R¹=OBzl: 63%; $[α]_D$ +38 (chloroform): Ref.: 38

α-L-Rhap-(1→4)-β-D-Glcp-(1→6)-D-Glc

β-Oleanolate; powder; m.p. 190—193°C; [α]_D +17.5 (methanol); Ref.: 74

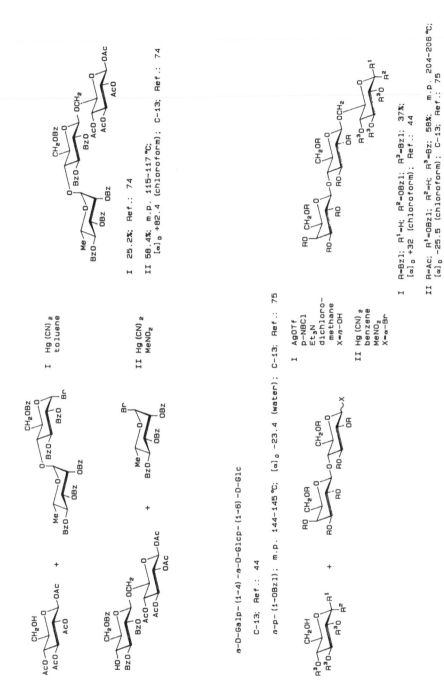

I   Hg(CN)₂
     toluene

II   Hg(CN)₂
     MeNO₂

I   25.2%; Ref.: 74

II   58.4%; m.p. 115—117°C;
     [α]_D +82.4 (chloroform); C-13: Ref.: 74

β-D-Galp-(1→4)-β-D-Glcp-(1→6)-D-Glc

C-13; Ref.: 44

β-p-(1→OBzl); m.p. 144—145°C; [α]_D -23.4 (water); C-13: Ref.: 75

I   AgOTf
     p-NBCl
     Et₃N
     dichloro-
     methane
     X=β-OH

II   Hg(CN)₂
     benzene
     MeNO₂
     X=α-Br

I   R=Bzl; R¹=H; R²=OBzl; R³=Bzl; 37%;
     [α]_D +32 (chloroform); Ref.: 44

II   R=Ac; R¹=OBzl; R²=H; R³=Bz; 58%; m.p. 204—206°C;
     [α]_D -25.5 (chloroform); C-13: Ref.: 75

β-D-Galp-(1→3)-β-D-GlcpNAc-(1→4)-D-Glc

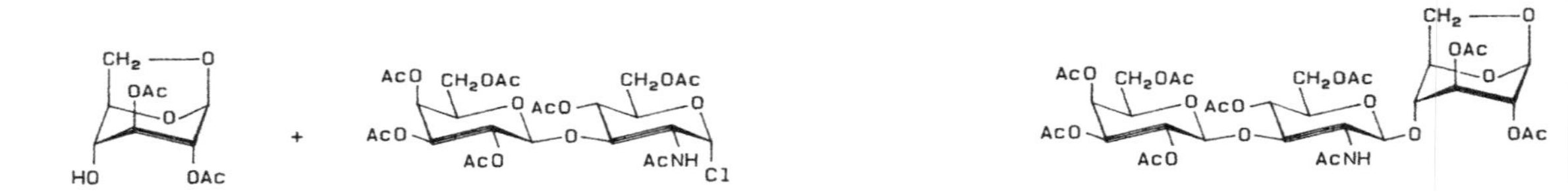

low yield: Ref.: 76

β-D-GlcpNAc-(1→4)-β-D-GlcpNAc-(1→6)-D-Glc

amorphous powder; [α]_D +1.4 (water); Ref.: 77

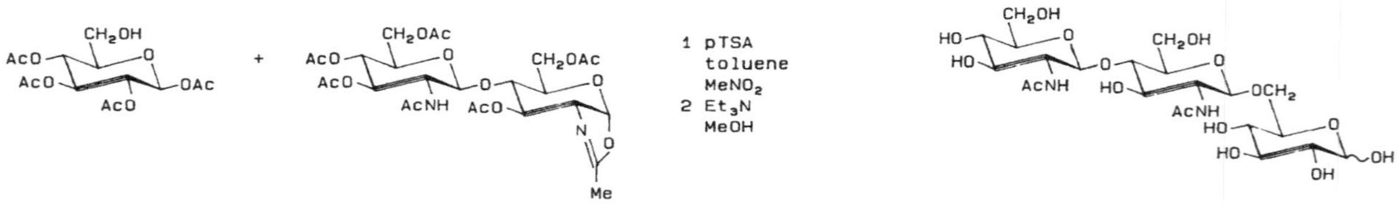

1 pTSA
  toluene
  MeNO_2
2 Et_3N
  MeOH

29%; Ref.: 77

a-D-Glcp-(1→6)-α-D-Manp-(1→2)-D-Glc

α-(1→1)-2,3-di-O-phytanyl-sn-glycerol

$[\alpha]_D$ +35.0 (chloroform): C-13; Ref.: 79

$[\alpha]_D$ +35 (chloroform-MeOH 1:1): C-13; Ref.: 78

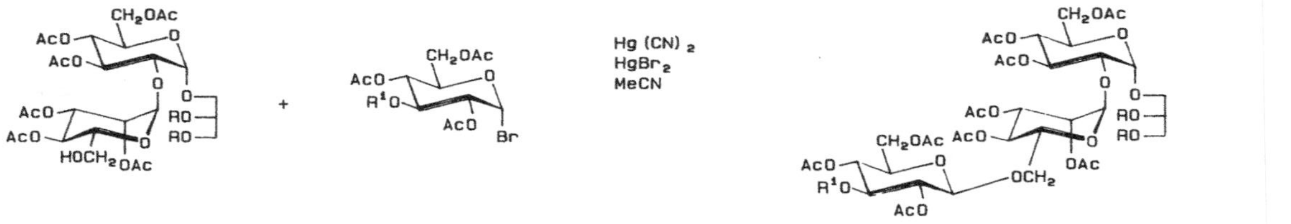

Hg(CN)$_2$
HgBr$_2$
MeCN

R=phytanyl: R^1=Ac; C-13; Ref.: 79

R=phytanyl; R^1=Ac; 74%;
$[\alpha]_D$ +49 (chloroform): C-13; Ref.: 78

R=phytanyl; R^1=CO-CH$_2$-CCl$_3$; 71%;
$[\alpha]_D$ +48 (chloroform): C-13; Ref.: 78

β-D-Galp-(1→6)-α-D-Manp-(1→2)-D-Glc

α-(1→1)-2,3-di-O-phytanyl-sn-glycerol

C-13; Ref.: 79

[α]$_D$ +32 (chloroform-MeOH 1:1); C-13; Ref.: 78

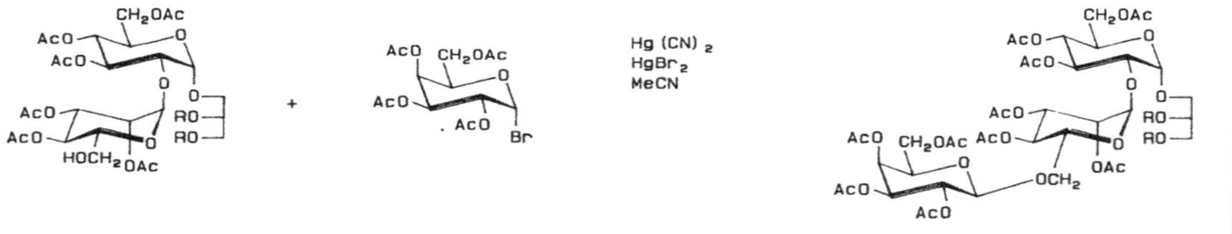

Hg(CN)$_2$
HgBr$_2$
MeCN

R=phytanyl; 74%; [α]$_D$ +43 (chloroform); C-13; Ref.: 78

α-D-Manp-(1→2)-α-D-Manp-(1→3)-D-Glc

α-p-(1-OMe); [α]$_D$ +110 (water); Ref.: 80

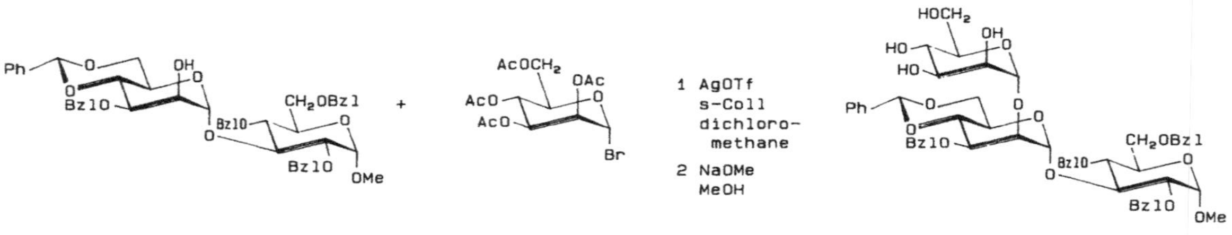

1 AgOTf
s-Coll
dichloro-
methane
2 NaOMe
MeOH

[α]$_D$ +63 (chloroform); Ref.: 80

α-D-Manp-(1→4)-α-L-Rhap-(1→3)-D-Glc

[α]$_D$ +22 (water); C-13; Ref.: 81

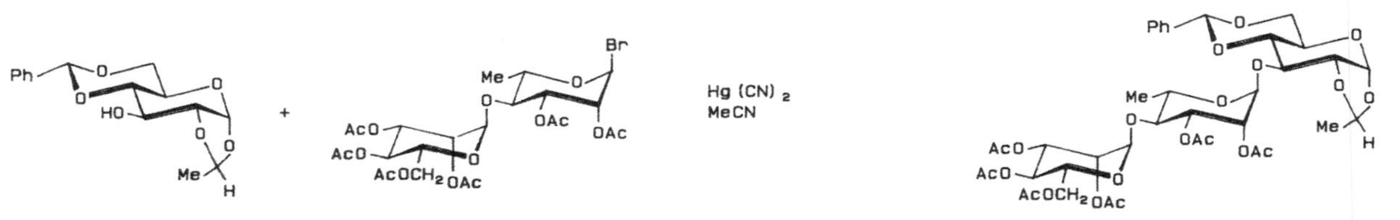

Hg(CN)$_2$
MeCN

85%; [α]$_D$ +11.5 (chloroform); Ref.: 81

α-L-Rhap-(1→3)-α-L-Rhap-(1→6)-D-Glc

Ref.: 82

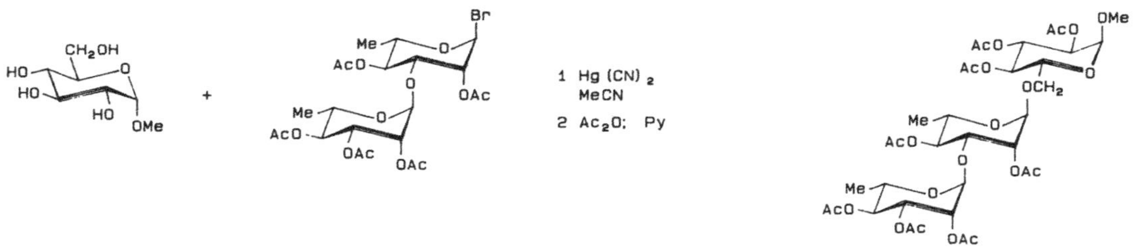

1 Hg(CN)$_2$
MeCN

2 Ac$_2$O; Py

α-D-Galp-(1→2)-α-D-Galp-(1→2)-D-Glc

α-acetate; thick syrup; $[\alpha]_D$ +149 (chloroform); C-13; Ref.: 83

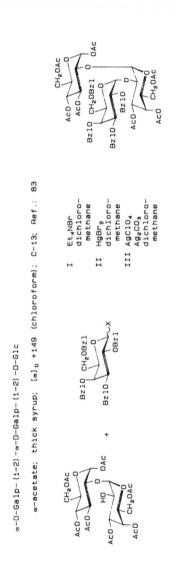

I   Et₄NBr dichloro-methane

II   HgBr₂ dichloro-methane

III   AgClO₄ Ag₂CO₃ dichloro-methane

I   X=α-Br: 13%

II   X=α-Br: 30%

III   X=Cl: 58%; $[\alpha]_D$ +91 (chloroform)

α-D-Glcp-(1→2)-α-D-Galp-(1→3)-D-Glc

α-p-(1→OMe); $[\alpha]_D$ +217 (water); C-13; Ref.: 84

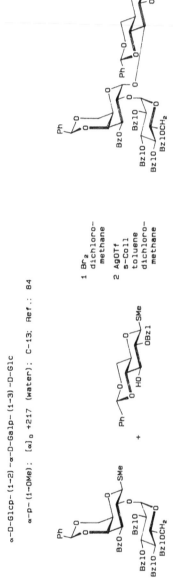

1 Br₂ dichloro-methane

2 AgOTf s-Coll toluene dichloro-methane

33%: $[\alpha]_D$ +109 (chloroform); C-13; Ref.: 84

α-D-Galp-(1→4)-α-D-Galp-(1→4)-D-Glc

34%; [α]$_D$ +122 (chloroform); C-13; Ref.: 85

AgOTf
TMU
dichloro-
methane

α-D-Galp-(1→6)-α-D-Galp-(1→6)-D-Glc

β-acetate; m.p. 106–107.5°C; [α]$_D$ +132.7 (chloroform); Ref.: 86

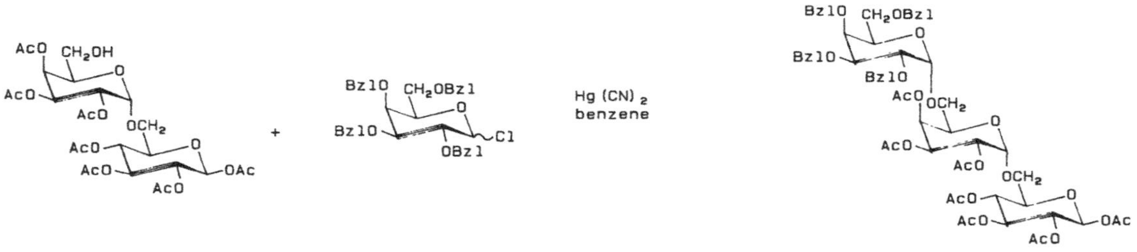

Hg(CN)$_2$
benzene

89%; Ref.: 86

α-D-Galp-(1→4)-β-D-Galp-(1→3)-D-Glc

acetate: m.p. 124 °C; [α]$_D$ +135 (chloroform); Ref.: 87

Hg(CN)$_2$
dichloro-
ethane

m.p. 105-108 °C; [α]$_D$ +110 (chloroform)

β-D-GalpNAc-(1→4)-β-D-Galp-(1→3)-D-Glc

hygroscopic powder: m.p. 175-178 °C; [α]$_D$ +24.7 (water); Ref.: 88

Hg(CN)$_2$
dichloro-
ethane

80%; m.p. 135-137 °C; [α]$_D$ +1.5

β-D-GlcpA-(1→3)-β-D-Galp-(1→4)-D-Glc

[α]$_D$ -13 (water): Ref.: 89

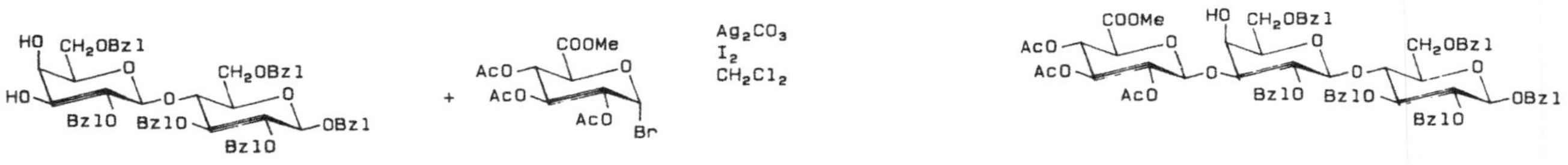

37%; m.p. 168 °C; [α]$_D$ -10.5 (chloroform)

α-D-GlcpN-(1→3)-β-D-Galp-(1→4)-D-Glc

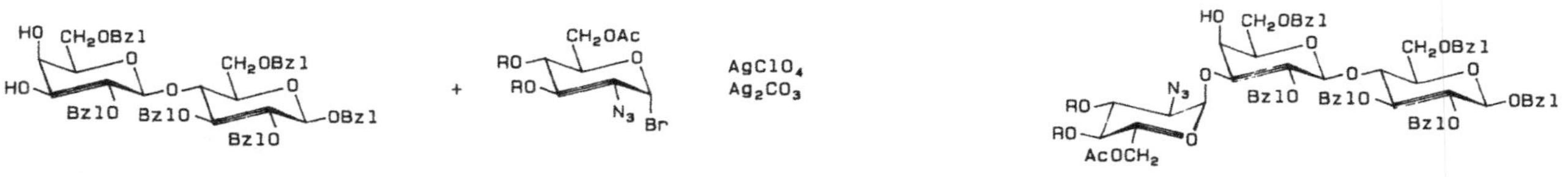

R=Ac: 29%; [α]$_D$ +61.2 (chloroform): Ref.: 90

R=Bzl: 23%; [α]$_D$ +33.4 (chloroform): Ref.: 90

α-D-GlcpN-(1→4)-β-D-Galp-(1→4)-D-Glc

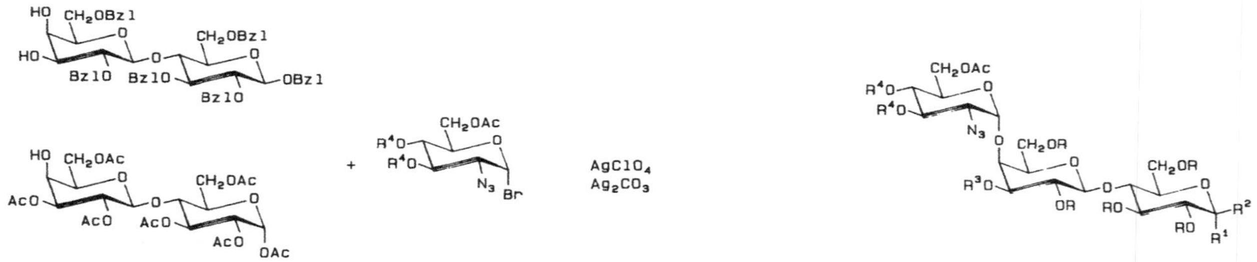

R=Bzl; $R^1=R^3=H$; $R^2=OBzl$; $R^4=Ac$: 34%; m.p. 123–125 °C; $[\alpha]_D$ +63.2 (chloroform): Ref.: 90

$R=R^4=Bzl$; $R^1=R^3=H$; $R^2=OBzl$: 20%; $[\alpha]_D$ +15.2 (chloroform): Ref.: 90

$R=R^3=Ac$; $R^1=OAc$; $R^2=H$; $R^4=Bzl$: 46%; $[\alpha]_D$ +74.04 (chloroform): Ref.: 90

β-D-GlcpN-(1-3)-β-D-Galp-(1-4)-D-Glc

β-D-GlcpNAc-(1-3)-β-D-Galp-(1-4)-D-Glc

    m.p. 205-209 °C; $[\alpha]_D$ +39.0 (water); Ref.: 90

    m.p. 205-209 °C; $[\alpha]_D$ +39.5 (water); Ref.: 91

β-D-(1-OMe); m.p. 244 °C (dec.); $[\alpha]_D$ +6 (water); Ref.: 92

I   AgOTf s-Coll MeNO₂

II AgSi dichloromethane

III Hg(CN)₂ dichloroethane

IV AgOTf s-Coll dichloromethane

V   pTSA dichloroethane

VI 1 BF₃.Et₂O dichloromethane 2 H⁺ R^2,R^3=PhCH

I   R=Bzl; R^1=Bzl; R^2=H; R^3=Bzl; R^4=Ac; R^5=NPhth; R^6=Ac
    83%; $[\alpha]_D$ +16.1 (chloroform); Ref.: 90, 93
    72%; $[\alpha]_D$ +20.6 (chloroform); Ref.: 94

II R=Bzl; R^1=Bzl; R^2=H; R^3=Bzl; R^4=Ac; R^5=N₃; R^6=Bzl
    9%; $[\alpha]_D$ +9.7 (chloroform); Ref.: 90

III R=Ac; R^1=Bzl; R^2=H; R^3=Ac; R^4=Bz; R^5=NHCOCHCl₂; R^6=Bz
    19%; m.p. 125-127 °C; $[\alpha]_D$ -50 (chloroform); Ref.: 91

IV R=Bz; R^1=Me; R^2,R^3=PhCH; R^4=Ac; R^5=NPhth; R^6=Ac
    44%; m.p. 279-281 °C; $[\alpha]_D$ +61 (chloroform); Ref.: 95

V   R=Bzl; R^1=Me; R^2=BzlpBr; R^3=BzlpBr; R^4=Ac; R^5=NHAc; R^6=Ac
    42%; m.p. 151-154 °C; $[\alpha]_D$ -20 (chloroform); Ref.: 92

VI R=Bz; R^1=Me; R^2=H; R^4=H; R^5=NPhth; R^6=Ac
    71%; m.p. 283 °C; $[\alpha]_D$ +56 (chloroform); Ref.: 95

VII R=Bzl; R^2=Bzl; R^3=Bzl; R^4=Bzl; R^5=NPhth; R^6=Ac
    $[\alpha]_D$ -2.2 (chloroform); Ref.: 96

VIII R=Bzl; R^2=Bzl; R^3=Bzl; R^4=Ac; R^5=NPhth; R^6=Ac
    95%; $[\alpha]_D$ -6.7 (chloroform); Ref.: 57

VII   MeOTf
      MeNO$_2$

VIII CuBr$_2$
      Et$_4$NBr
      AgOTf
      MeNO$_2$

β-D-GlcpN-(1→4)-β-D-Galp-(1→4)-D-Glc

β-D-GlcpNAc-(1→4)-β-D-Galp-(1→4)-D-Glc

syrup: [α]$_D$ +6.3 (water): Ref.: 97

I   Ag$_2$CO$_3$
     AgClO$_4$
     dichloro-
     methane

II   AgOTf
     s-Coll
     MeNO$_2$

III AgSi
     dichloro-
     methane

III

I   R^1=Me: R=R^2=R^3=Bzl; 54%;
     [α]$_D$ +25.0 (dichloromethane): C-13; Ref.: 97

     R^1=Me: R=R^2=Bzl; R^3=Bz; 8%;
     [α]$_D$ +28.2 (dichloromethane): C-13; Ref.: 97

II   R=R^1=R^3=Bzl; R^2=H; 23%;
     [α]$_D$ +3.3 (chloroform): Ref.: 94

III R=R^1=R^3=Bzl; R^2=H; 86%;
     [α]$_D$ +18.3 (chloroform): Ref.: 90, 93

III R^1=R=R^3=Bzl; R^2=H; 83%;
     [α]$_D$ +12.3 (chloroform): Ref.: 90

β-D-GlcpNAc-(1→6)-β-D-Galp-(1→4)-D-Glc

β-D-(1→OBzl): hygroscopic amorphous powder; [α]_D -6.8 (methanol): Ref.: 99

β-D-(1→OBzl): amorphous; [α]_D -16 (water): Ref.: 98

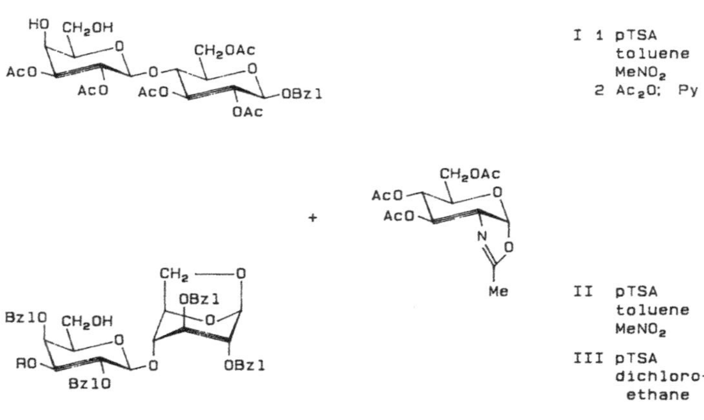

I   1 pTSA
      toluene
      MeNO_2
    2 Ac_2O; Py

II pTSA
    toluene
    MeNO_2

III pTSA
    dichloro-
    ethane

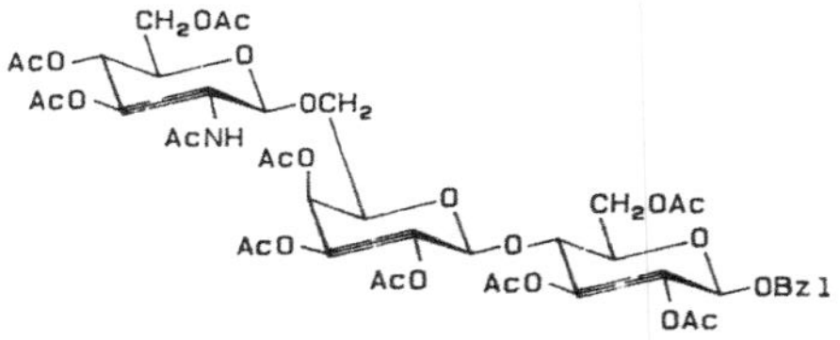

I   82%; m.p. 152-155 °C; [α]_D -33 (chloroform): Ref.: 98

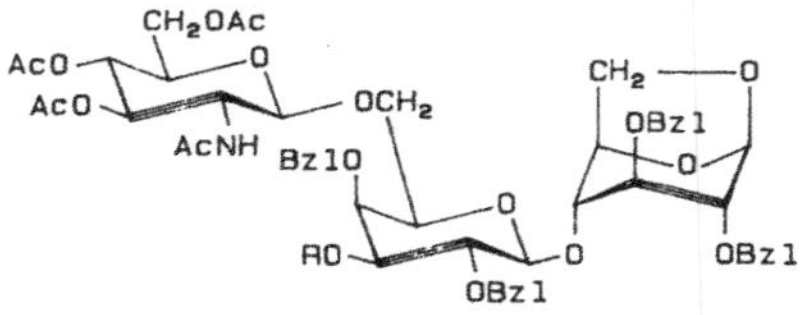

II R=Bzl; 58.5%; [α]_D -13.1 (chloroform): Ref.: 99

III R=H; 16%; [α]_D -16 (chloroform): Ref.: 100

α-D-Galp-(1-4)-β-D-Galp-(1-4)-D-Glc

[α]_D +101 (water): Ref.: 101

β-D-(1-OMe): amorphous; [α]_D +65 (water); C-13: Ref.: 102

α-D-(1-OMe): colourless glass; [α]_D +63 (water); C-13; Ref.: 103

β-D-(1-OEt): [α]_D +71 (water); C-13: Ref.: 85

I   Et₄NBr
    DMF
    dichloro-
    methane

II AgOTf
    s-Coll
    dichloro-
    ethane

III 1 AgOTf
      s-Coll
      toluene

    2 H₂: PdC

I   R=Bz; R¹=H; R²=OMe; R³=Bz; X=Br; 34%;
     colourless oil; C-13; Ref.: 103

II   R=Bz; R¹=OBz; R²=H; R³=Bz; X=Cl; 89%;
     [α]_D +93 (chloroform); C-13; Ref.: 101

     R=Bz; R¹=H; R²=OBz; R³=Bz; X=Cl; 83%;
     [α]_D +48 (chloroform); C-13; Ref.: 101

III R=Bz; R¹=OBz; R²=H; R³=Bz; X=Br; 64%;
     [α]_D +124 (chloroform); C-13; Ref.: 102

BzlO CH₂OAc
R⁵O
BzlO

\+

BzlO CH₂OAc
AllO
BzlO

Br

Br

IV Ag₂CO₃
    AgClO₄
    dichloro-
    methane

IV

BzlO CH₂OAc
BzlO
CH₂OBz

O CH₂OAc
AcO AcO

CH₂OAc
AcO AcO R¹
         R²

IV R=Ac; R¹=OAc; R²=H; R³=Bz; R⁵=Bzl; 57%;
    [α]_D +30.2 (chloroform); Ref.: 104

    R=Ac; R¹=OAc; R²=H; R³=Bz; R⁵=All; 47%;
    [α]_D +36.5 (chloroform); Ref.: 104

BzlO CH₂OAc
TCAO
BzlO

Br

\+

IV

BzlO CH₂OAc
TCAO
CH₂OBzl

O CH₂OBz
BzlO BzlO

CH₂OBzl
BzlO R¹
      R²

IV R=Bzl; R¹=H; R²=OBzl; R³=Bz; 18%;
    [α]_D +30.8 (chloroform); Ref.: 104

AcO CH₂OAc
AcO

O CH₂OAc
AcO X

\+

CH₂
    O
OR O OR
HO

V   Hg(CN)₂
    dichloro-
    ethane

VI 1 Hg(CN)₂
      Et₂O; MeCN
      dichloro-
      methane

    2 MeOH,
      Ba(OMe)₂

    3 Ac₂O

AcO CH₂OAc
AcO

O CH₂OAc
AcO AcO

CH₂
    O
    OAc
    OAc

V   R=Ac; X=Br; m.p. 118 °C;
    [α]_D +36.5 (chloroform); Ref.: 87

VI R=Bz; X=Cl; m.p. 118 °C;
    [α]_D +38.5 (chloroform); Ref.: 87

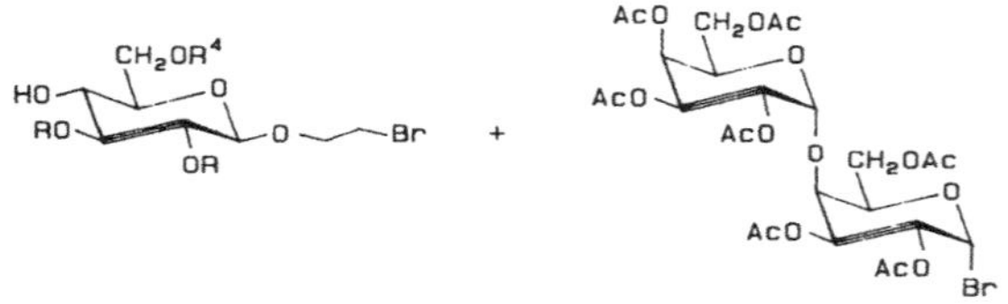

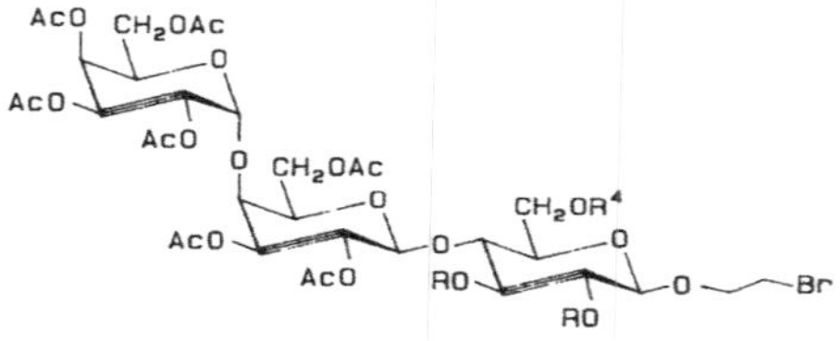

VII R=Bzl, R⁴=Bzl; 70%; m.p. 155-158 °C;
$[\alpha]_D$ +58 (chloroform); C-13; Ref.: 85

R=Bz; R⁴=Bzl; 11%; m.p. 189-192 °C;
$[\alpha]_D$ +65 (chloroform); C-13; Ref.: 85

VII AgOTf
TMU
dichloro-
methane

β-D-Galp-(1→3)-β-D-Galp-(1→4)-D-Glc

$[\alpha]_D$ +26.1 (water); Ref.: 105

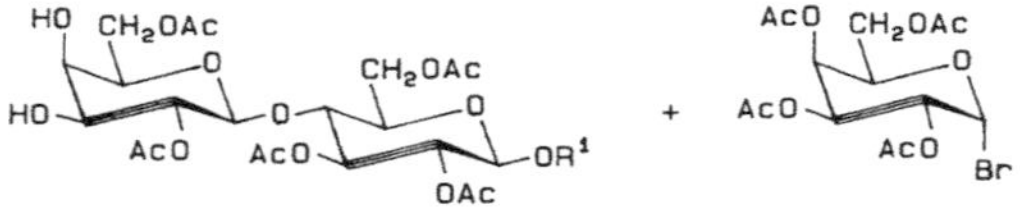

I Hg(CN)₂
benzene
MeNO₂

II 1 Hg(CN)₂
benzene
MeNO₂
2 NaOMe

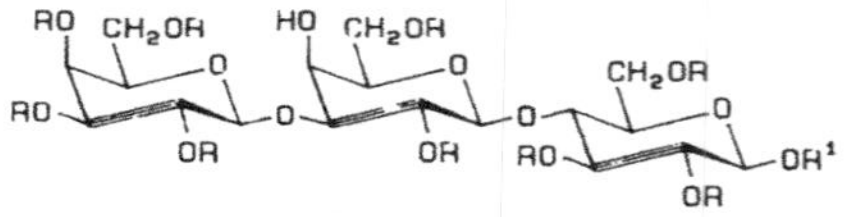

R=Ac; R¹=Bzl; Ref.: 76

I R=Ac; R¹=Bzl; 50%; $[\alpha]_D$ -5.2 (chloroform); Ref.: 105

II R=H; R¹=-CH₂-CH(NH-CO-C₁₅H₃₁)-CH(OH)-(CH₂)₁₄-CH₃;
20%; m.p. 257-258 °C; $[\alpha]_D$ +5.7 (Py); Ref.: 106

β-D-Galp-(1→6)-β-D-Galp-(1→4)-D-Glc

white powder; [α]$_D$ +36 (water); Ref.: 107

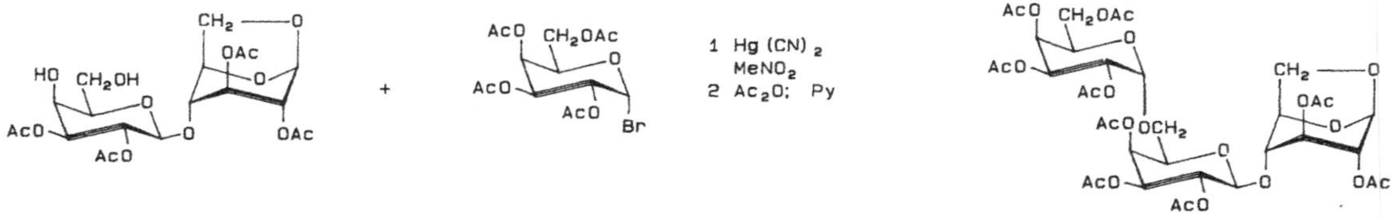

+

1 Hg(CN)$_2$
  MeNO$_2$
2 Ac$_2$O; Py

62%; m.p. 105—107°C; [α]$_D$ −41.5 (chloroform); Ref.: 107

β-D-GalpN-(1→3)-β-D-Galp-(1→4)-D-Glc

β-D-GalpNAc-(1→3)-β-D-Galp-(1→4)-D-Glc

m.p. 148°C; [α]$_D$ −7 (methanol); Ref.: 108

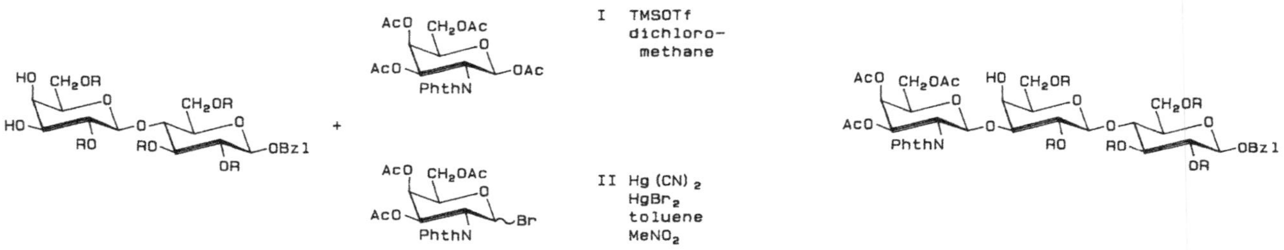

+

I TMSOTf
   dichloro-
   methane

II Hg(CN)$_2$
   HgBr$_2$
   toluene
   MeNO$_2$

I R=Ac; 53%; [α]$_D$ −8 (methanol); Ref.: 93, 108

II R=Bzl; 42.5%; m.p. 121°C; [α]$_D$ −32 (acetone); Ref.: 108

β-D-GalpN-(1→4)-β-D-Galp-(1→4)-D-Glc

β-D-GalpNAc-(1→4)-β-D-Galp-(1→4)-D-Glc

hygroscopic powder; m.p. 185—188 °C; $[\alpha]_D$ +30.3 (water): Ref.: 88

$[\alpha]_D$ +41 (methanol): Ref.: 93, 108

β-p-(1-OMe): $[\alpha]_D$ -10.1 (methanol); C-13: Ref.: 97

β-p-(1-OMCO): $[\alpha]_D$ -0.1 (methanol); C-13: Ref.: 109

β-p-$\begin{array}{c}\text{HN-CO-C}_{23}\text{H}_{47}\\ \text{O-}\overset{|}{\text{C}}_{13}\text{H}_{27}\\ \overset{|}{\text{OH}}\end{array}$; $[\alpha]_D$ -2.8 (chloroform—methanol 1:1): Ref.: 110

R=Bz; R^1=R^2=Ac; R^3=NHCOCHCl$_2$: Ref.: 76

I   R=R^1=R^2=Ac; R^3=NHAc: Ref.: 111

R=R^1=R^2=Ac; R^3=NHAc: 20%: Ref.: 76

II R=R^1=R^2=Ac; R^3=NHAc: 57%: m.p. 123—125 °C; $[\alpha]_D$ -35.2: Ref.: 88

R=Ac; R^1=Bz; R^2=H; R^3=NHAc: m.p. 139—141 °C; $[\alpha]_D$ -5.3: Ref.: 88

I   Hg(CN)$_2$

II Hg(CN)$_2$
    dichloro-
    ethane

III AgOTf
    s-Coll
    MeNO$_2$

III 65%; $[\alpha]_D$ -19.9 (chloroform); C-13: Ref.: 109

IV   Hg(CN)$_2$
     HgBr$_2$
     toluene, MeNO$_2$

V   AgOTf
     dichloro-
     ethane

VI   Ag$_2$CO$_3$
     AgClO$_4$
     dichloro-
     methane

VII 1 Ag$_2$CO$_3$
       dichloro-
       methane
     2 Ac$_2$O; Py

IV   R=Ac; R^3=NPhth; R^4=Bzl; R^5=H; 41.3%; m.p. 127 °C;
     [α]$_D$ −47 (chloroform); Ref.: 108

V   R=Ac; R^3=NPhth; R^4=Bzl; R^5=Bzl; 80%;
     [α]$_D$ +2.3 (chloroform); Ref.: 110

VI   R=Bz; R^3=NPhth; R^4=Me; R^5=Bzl; 48.5%;
     [α]$_D$ +31.5 (dichloromethane); C-13; Ref.: 97

VII R=Ac; R^3=N$_3$; R^4=Bzl; R^5=Ac; 72%;
     [α]$_D$ +13 (acetone); Ref.: 93, 108

α-L-Fucp-(1→2)-β-D-Galp-(1→4)-D-Glc

white solid; m.p. 230—231°C; [α]$_D$ −48.6 → −50.2 (water); Ref.: 112

[α]$_D$ −43 → −48 (water); C-13; Ref.: 113

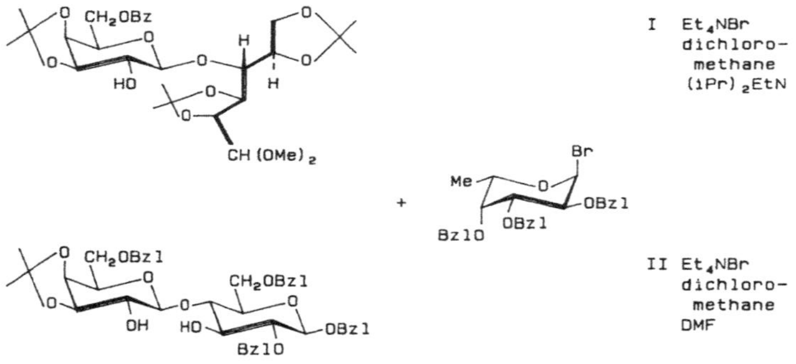

I Et$_4$NBr
   dichloro-
   methane
   (iPr)$_2$EtN

II Et$_4$NBr
   dichloro-
   methane
   DMF

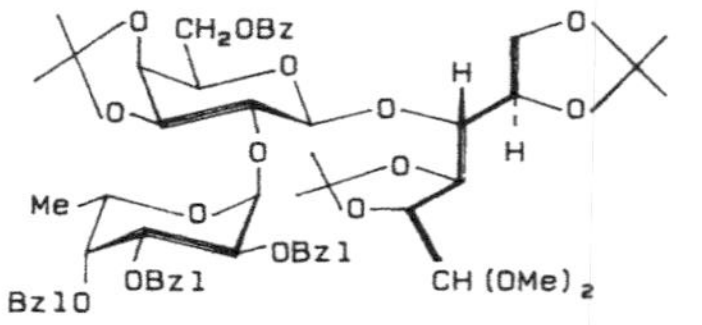

I 50.3%; [α]$_D$ −54.1 (chloroform); Ref.: 112

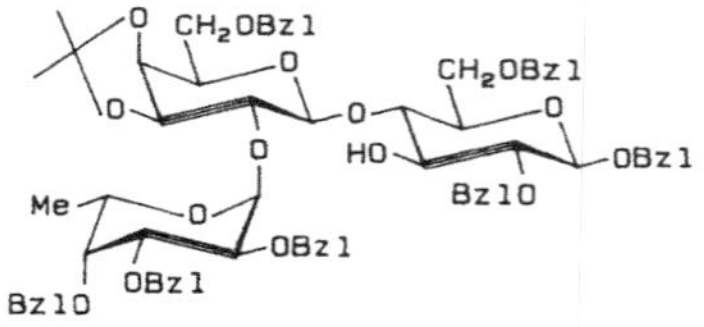

II 55%; [α]$_D$ −47 (chloroform); C-13; Ref.: 113

α-L-Fucp-(1→3)-β-D-Galp-(1→4)-D-Glc

white powder: [α]_D −39 → −42.6 (water); Ref.: 114

m.p. 255−257 °C (dec.); [α]_D −25.3 → −30 (water); C−13; Ref.: 115

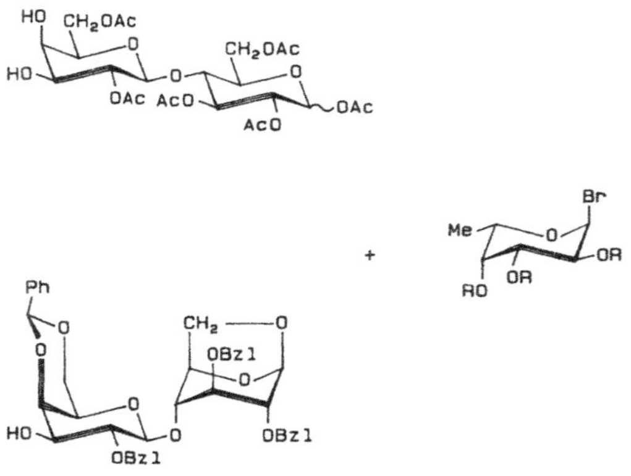

I   Et₄NBr
    dichloro-
    methane
    (iPr)₂EtN

II Hg(CN)₂
    benzene
    MeNO₂

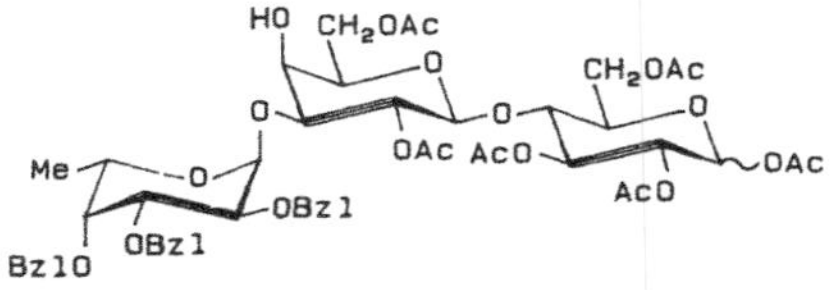

I   R=Bzl; 42.5%; m.p. 156−158 °C (softening at 125−130 °C);
    [α]_D −20.5 (chloroform); Ref.: 114

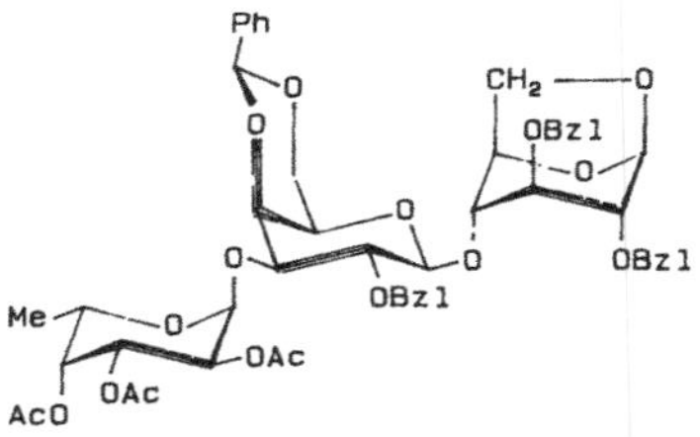

II R=Ac; 37.1%; [α]_D −66.8 (chloroform); Ref.: 115

α-L-Fucp-(1→4)-β-D-Galp-(1→4)-D-Glc

hygroscopic amorphous powder;   [α]$_D$ -37.2 (water);   C-13;   Ref.: 116

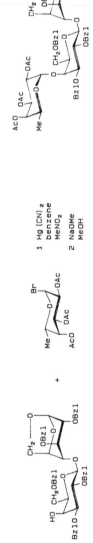

1 Hg(CN)$_2$
   benzene
   MeNO$_2$

2 NaOMe
   MeOH

13.4%;   [α]$_D$ -52.5 (chloroform);   Ref.: 116

α-L-Fucp-(1→6)-α-D-Galp-(1→4)-D-Glc

white hygroscopic solid: [α]_D -28.3 (methanol), -23 → -21 (water); Ref.: 117

hygroscopic glass; [α]_D -27.8 (water); C-13; Ref.: 116

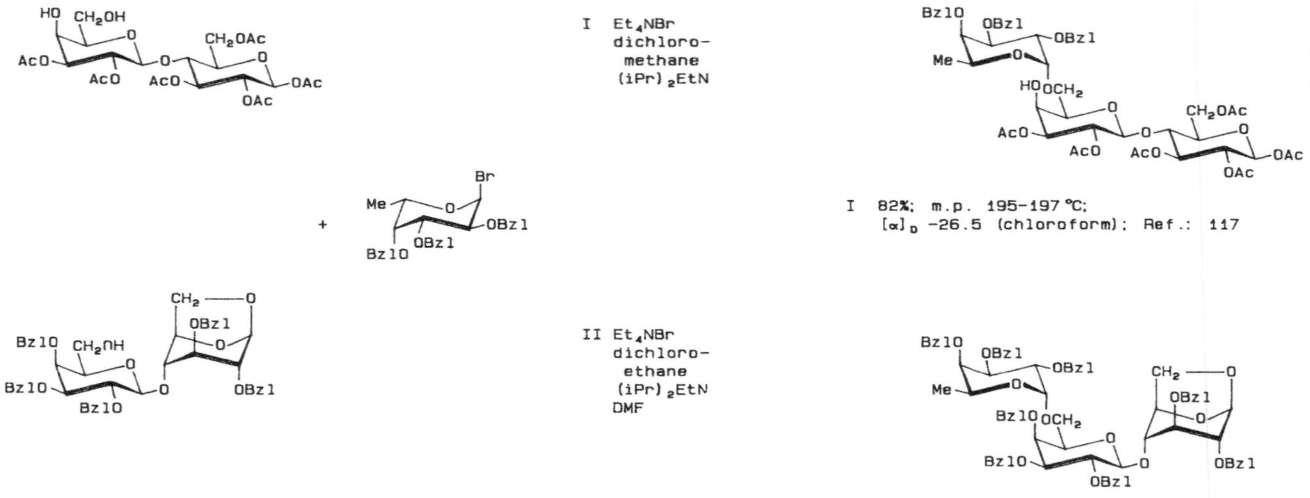

I Et₄NBr
   dichloro-
   methane
   (iPr)₂EtN

II Et₄NBr
   dichloro-
   ethane
   (iPr)₂EtN
   DMF

I 82%; m.p. 195-197 °C;
   [α]_D -26.5 (chloroform); Ref.: 117

II 51.5%; [α]_D -49.7 (chloroform); Ref.: 116

β-L-Fucp-(1→3)-β-D-Galp-(1→4)-D-Glc

[α]$_D$ +54.2 → +51 (water): Ref.: 118

m.p. 265-267 °C (dec.): [α]$_D$ +67.1 (water): C-13: Ref.: 115

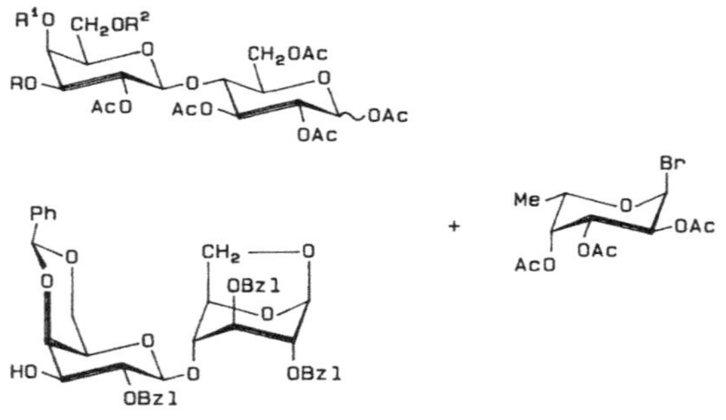

I   Ag$_2$CO$_3$
     CHCl$_3$

II   Ag$_2$CO$_3$
     CH$_2$Cl$_2$

III Hg(CN)$_2$
     benzene
     MeNO$_2$

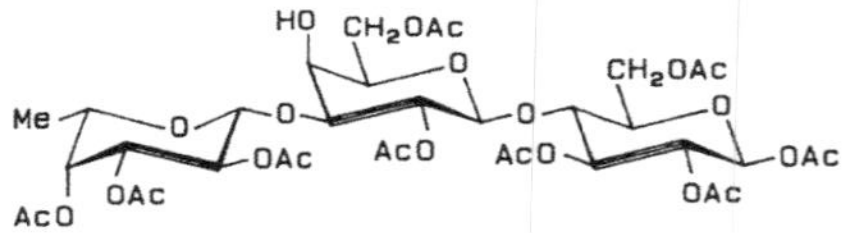

I   R=Ac; R^1=R^2=H; β-OAc; 27%; m.p. 222-224 °C;
     [α]$_D$ +7.7 (chloroform): Ref.: 118

II   R=R^1=H; R^2=Ac; m.p. 222-224 °C; Ref.: 114

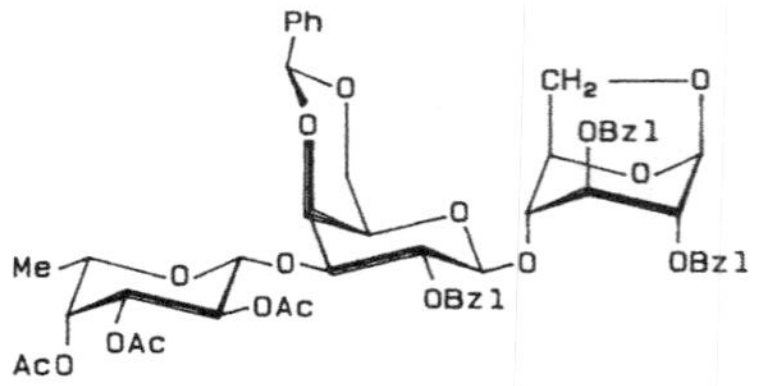

III 43%; m.p. 106-108 °C;
     [α]$_D$ +1.3 (chloroform): Ref.: 115

β-L-Fucp-(1→4)-β-D-Galp-(1→4)-D-Glc

hygroscopic amorphous powder; [α]$_D$ +36.8 (water); C-13; Ref.: 116

1 Hg(CN)$_2$
benzene
MeNO$_2$

2 NaOMe
MeOH

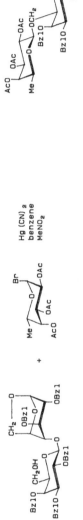

25.6%; [α]$_D$ -10.4 (chloroform); Ref.: 116

β-L-Fucp-(1→6)-β-D-Galp-(1→4)-D-Glc

hygroscopic amorphous powder; [α]$_D$ +48.4 (water); C-13; Ref.: 116

Hg(CN)$_2$
benzene
MeNO$_2$

amorphous powder

α-D-Neup5Ac-(2→3)-β-D-Galp-(1→4)-D-Glc

[α]$_D$ +19.2 (water): Ref.: 119

amorphous: [α]$_D$ +21.3 (water): Ref.: 120

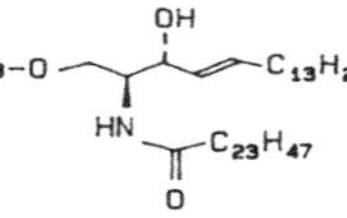

C$_{13}$H$_{27}$: [α]$_D$ +4.1 (chloroform-methanol 1:1): Ref.: 121

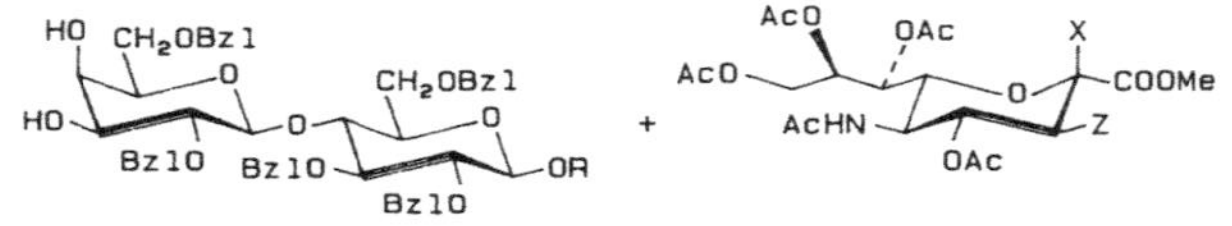

I   Hg(CN)$_2$: HgBr$_2$
     dichloroethane

II   Ag$_2$CO$_3$: AgClO$_4$
     toluene
     dichloromethane

III AgOTf
     toluene
     dichloroethane

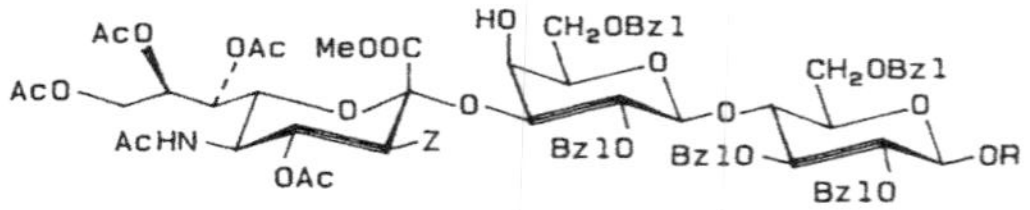

I   R=Bzl; Z=H; X=Cl;
     [α]$_D$ +5.8 (chloroform): Ref.: 119, 121

II   R=Bzl; Z=H; X=Br; 15.6%;
     [α]$_D$ -4.8 (chloroform): Ref.: 120

III R=All; Z=OH; X=Br; 24%;
     [α]$_D$ -4.4 (chloroform): Ref.: 122

α-D-Neup5Ac-(2→6)-β-D-Galp-(1→4)-D-Glc

amorphous powder; [α]_D +5.6 (water); C-13; Ref.: 123

Hg(CN)₂
HgBr₂
dichloro-
methane

21%; [α]_D -30.2 (chloroform); Ref.: 123

β-D-Neup5Ac-(2→3)-β-D-Galp-(1→4)-D-Glc

[α]$_D$ +11.6 (water); Ref.: 119

amorphous: [α]$_D$ +12.3 (water); Ref.: 120

[α]$_D$ -7.7 (chloroform-methanol 1:1); Ref.: 121

I   Hg(CN)$_2$: HgBr$_2$
    dichloroethane

II   Ag$_2$CO$_3$: AgClO$_4$
    toluene
    dichloromethane

III   AgOTf
    toluene
    dichloroethane

I   R=Bzl; R^1=Bzl; Z=H; X=Cl;
    [α]$_D$ +3.1 (chloroform); Ref.: 119

    R=Ac; R^1=All; Z=H; X=Cl; 6%;
    [α]$_D$ -8.9 (chloroform); Ref.: 121

II   R=Bzl; R^1=Bzl; Z=H; X=Br; 18.7%;
    [α]$_D$ -2.8 (chloroform); Ref.: 120

III   R=Bzl; R^1=All; Z=OH; X=Br; 28%;
    [α]$_D$ -5.8 (chloroform); Ref.: 122

β-D-Neup5Ac-(2→6)-β-D-Galp-(1→4)-D-Glc

amorphous powder; [α]_D +1.8 (water); C-13; Ref.: 123

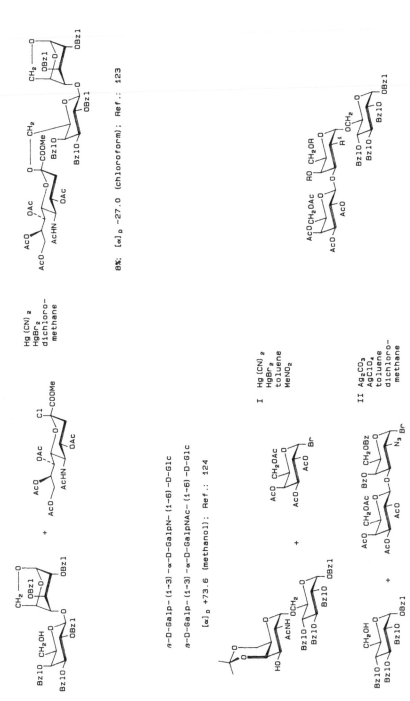

8%; [α]_D -27.0 (chloroform); Ref.: 123

β-D-Galp-(1→3)-α-D-GalpN-(1→6)-D-Glc

β-D-Galp-(1→3)-α-D-GalpNAc-(1→6)-D-Glc

[α]_D +73.6 (methanol); Ref.: 124

I   R, R= =C(Me)₂; R¹=NHAc; 74.9%; m.p. 56 °C;
    [α]_D +34.1 (chloroform); Ref.: 124

II R=Bz; R¹=N₃; 85%; [α]_D +38 (acetone); Ref.: 125

β-D-Galp-(1→3)-β-D-GalpNAc-(1→4)-D-Glc

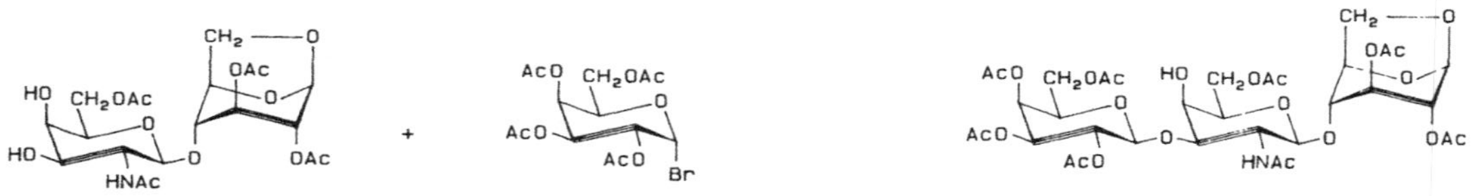

Ref.: 76

α-L-Fucp-(1→4)-α-L-Fucp-(1→3)-D-Glc

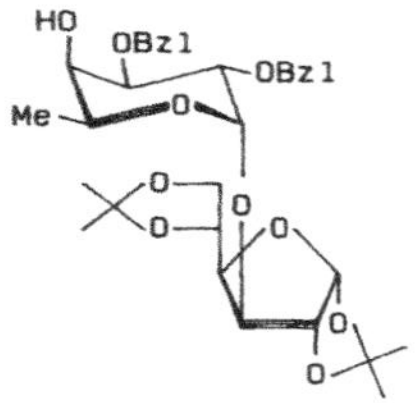

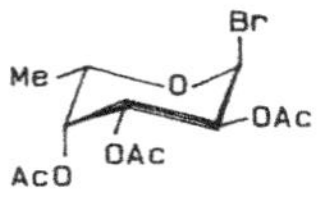

I 1 Hg(CN)₂
      benzene
      MeNO₂
   2 NaOMe

II 1 Hg(CN)₂
      MeCN
   2 NaOMe

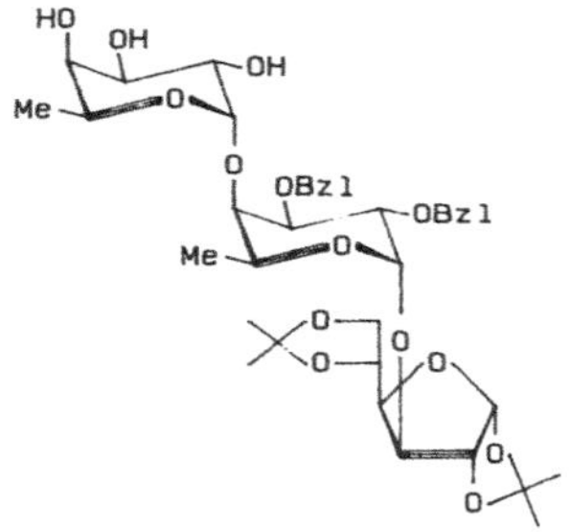

I

II

+

I 20%; [α]_D −106.5 (chloroform); Ref.: 126
II 4.4%; Ref.: 126

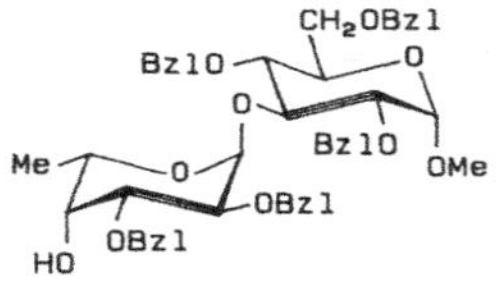

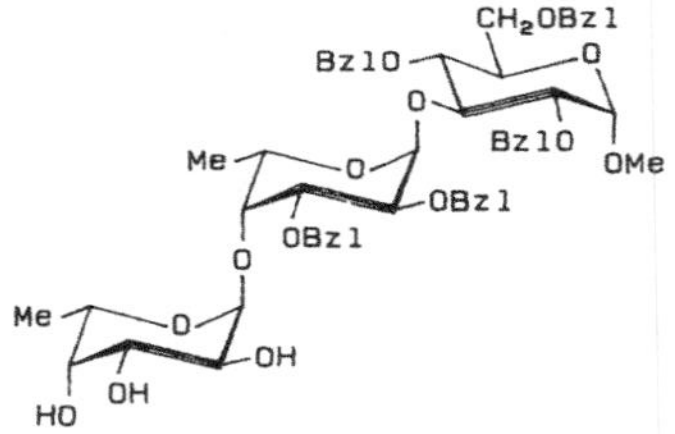

I 11%; [α]_D −86 (chloroform); Ref.: 126

β-L-Fucp-(1→4)-α-L-Fucp-(1→3)-D-Glc

[α]$_D$ -60 (methanol); C-13: Ref.: 126

α-p-(1→OMe); m.p. 223-226°C; [α]$_D$ -24 (methanol); C-13: Ref.: 126

I   1 Hg(CN)$_2$
        benzene
        MeNO$_2$
     2 NaOMe

II 1 Hg(CN)$_2$
        MeCN
     2 NaOMe

I

+

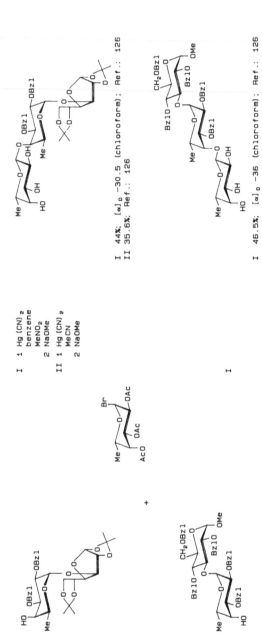

I   44%: [α]$_D$ -30.5 (chloroform); Ref.: 126
II 35.6%: Ref.: 126

I   46.5%: [α]$_D$ -36 (chloroform); Ref.: 126

α-D-Galp-(1→6)
                    \
                     α-D-Fruf
α-D-Glcp-(1→2)
                    /

m.p. 120-123 °C; [α]$_D$ +122.5 (water): Ref.: 127

+

Hg(CN)$_2$
benzene

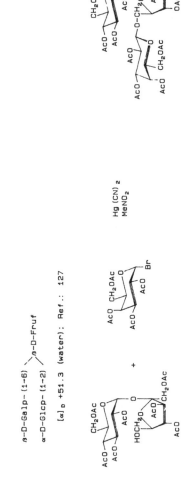

59%; [α]$_D$ +53.5 (chloroform): Ref.: 127

β-D-Galp-(1→6)
                    \
                     α-D-Fruf
α-D-Glcp-(1→2)
                    /

[α]$_D$ +51.3 (water): Ref.: 127

+

Hg(CN)$_2$
MeNO$_2$

22%; m.p. 87-89 °C; [α]$_D$ +44.9 (chloroform): Ref.: 127

β-D-Neup5Ac-(2→8)-α-D-Neup5Ac-(2→6)-D-Glc

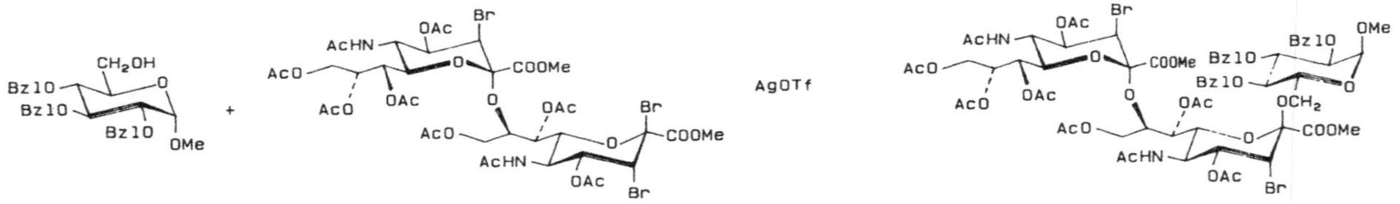

[α]$_D$ +31.7 (chloroform): Ref.: 128

α-D-Glcp-(1→4)
α-L-Araf-(1→6)   D-Glc

β-D-(1→OMe): m.p. 102—104 °C: [α]$_D$ +5.1 (water): C-13: Ref.: 29

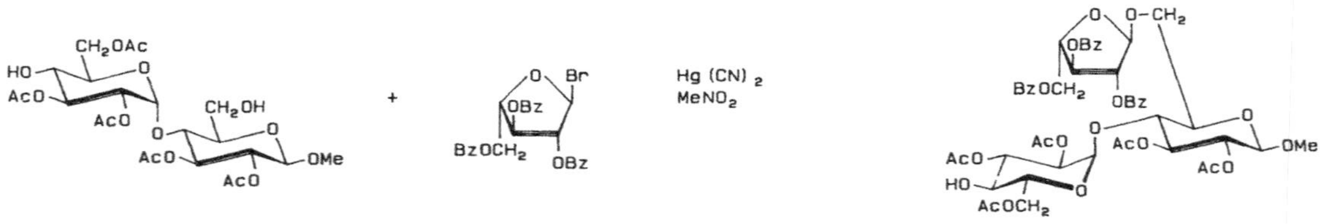

62.2%: m.p. 86—87 °C: [α]$_D$ +16.8 (chloroform): C-13: Ref.: 29

α-D-Glcp-(1→2)
                   D-Glc
α-D-Glcp-(1→4)

$[\alpha]_D$ +142.3 (water); Ref.: 26

Hg(CN)$_2$
HgBr$_2$
MeCN

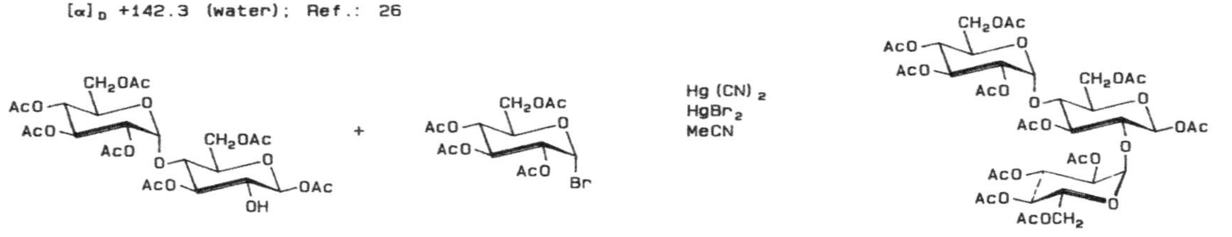

21%; $[\alpha]_D$ +103.0 (chloroform); Ref.: 26

α-D-Glcp-(1→2)
                   D-Glc
α-D-Glcp-(1→6)

Et$_2$O

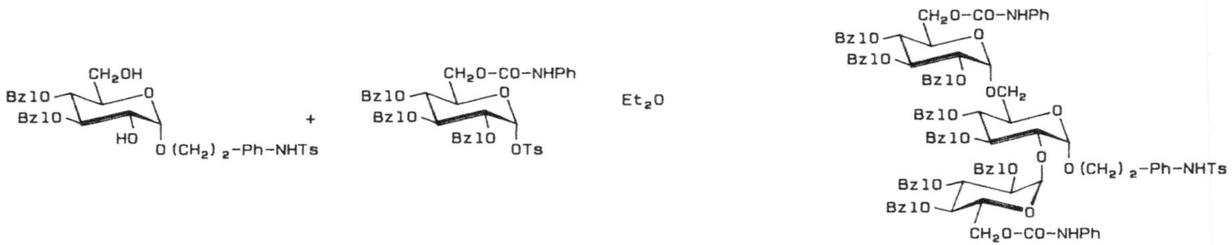

60%; $[\alpha]_D$ +77 (chloroform); C-13; Ref.: 129

α-D-Glcp-(1-3)
$\qquad\qquad\qquad$ D-Glc
α-D-Glcp-(1-6)

foam; $[\alpha]_D$ +126 (water); C-13; Ref.: 130

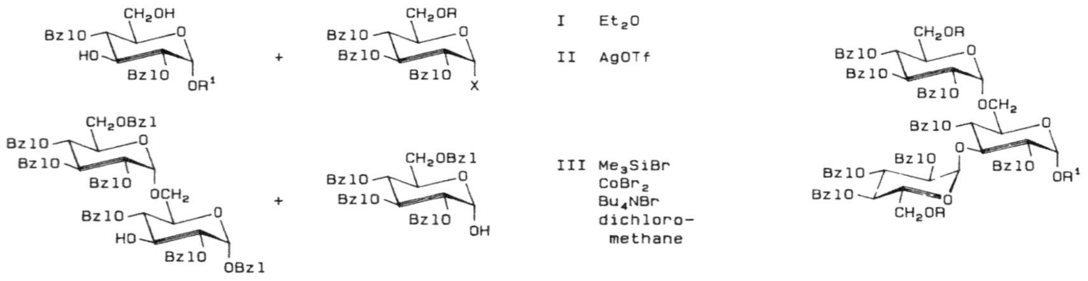

I   Et₂O
     AgOTf

III Me₃SiBr
     CoBr₂
     Bu₄NBr
     dichloro-
     methane

I   R¹=(CH₂)₂-Ph-NHTs; X=OTs; R=CO-NH-Ph; 61%;
     $[\alpha]_D$ +81.8 (chloroform); C-13; Ref.: 129

II   R¹=Bzl; R=Ac; X=Cl; $[\alpha]_D$ +82.2 (CDCl₃); C-13; Ref.: 131

III R¹=R=Bzl; 19%; $[\alpha]_D$ +79 (chloroform); C-13; Ref.: 130

α-D-Glcp-(1→3)⟍
                 ⟩D-Glc
β-D-Glcp-(1→6)⟋

AgOTf
p-NBCl
Et₃N
dichloro-
methane

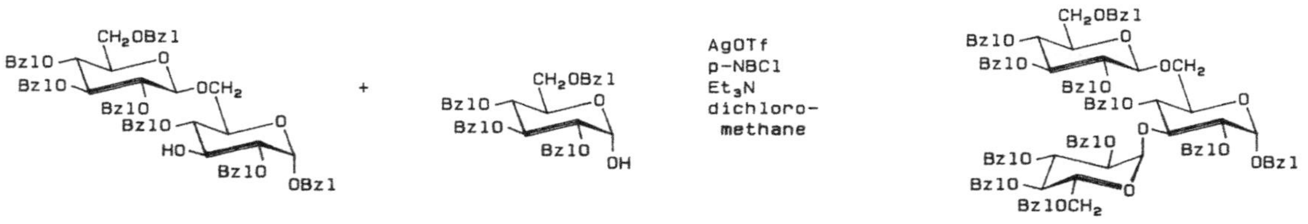

19%; [α]$_D$ +59 (chloroform); C-13; Ref.: 132

α-D-Glcp-(1-4)\
                  D-Glc
α-D-Glcp-(1-6)/

syrup: [α]_D +125 (water); Ref.: 133

[α]_D +136.5 (water); Ref.: 129

[α]_D +139 (water); C-13; Ref.: 134

foam: [α]_D +139 (water); C-13; Ref.: 135

β-D-(1-OMe); syrup: [α]_D +85.4 (water); C-13; Ref.: 136

I title compound

R¹=H; R²=OAc; Ref.: 133
I 1 "Koenigs-Knorr"
  2 hydrogenation
  3 NaOMe

II Et₄NBr
   DMF
   dichloromethane

III AgOTf
    pNBC1
    DMA

IV Et₂O

II R=Ac; R¹=H; R²=OMe; 48%;
   [α]_D +55.2 (chloroform); C-13; Ref.: 136

III R=Bzl; R¹=OBzl; R²=H; 62%;
    [α]_D +82 (chloroform); C-13; Ref.: 134, 135

IV R³=(CH₂)₂-Ph-NHTs; 67%;
   [α]_D +63.9 (chloroform); C-13; Ref.: 129
   R³=Me; 83%; [α]_D +59.2 (chloroform); C-13; Ref.: 129

β-D-Glcp-(1→2)
α-D-Glcp-(1→4)   D-Glc

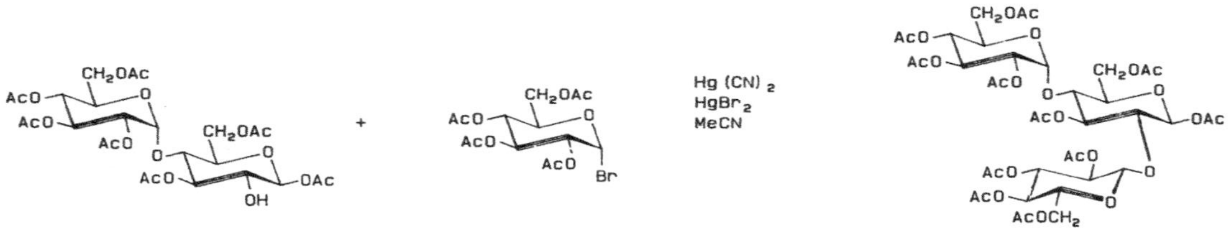

Hg(CN)₂
HgBr₂
MeCN

14%; m.p. 195—196 °C; $[\alpha]_D$ +46.5 (chloroform); Ref.: 26

α-D-Glcp-(1→4)
β-D-Glcp-(1→6)   D-Glc

$[\alpha]_D$ +82; Ref.: 137

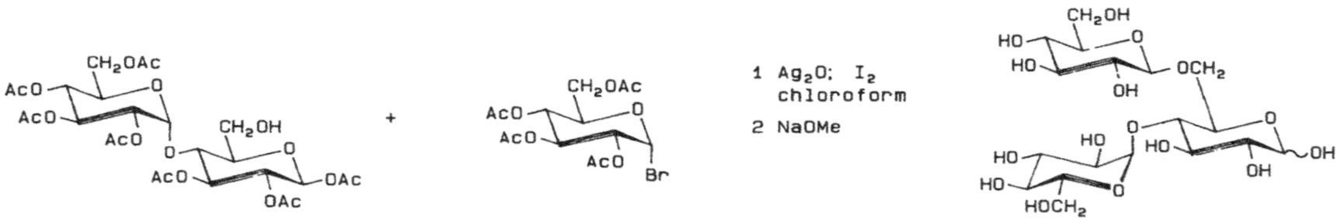

1 Ag₂O; I₂
  chloroform
2 NaOMe

α-D-Glcp-(1→4) \
        D-Glc
α-D-Galp-(1→6) /

hygroscopic amorphous powder; $[\alpha]_D$ +155 (methanol); Ref.: 138

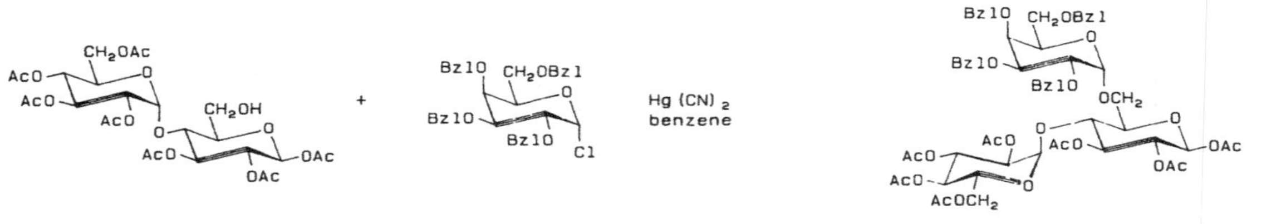

Hg(CN)₂
benzene

37%; $[\alpha]_D$ +55.8 (chloroform); Ref.: 138

α-D-Glcp-(1→4) \
        D-Glc
β-D-Galp-(1→6) /

m.p. 168-170 °C; $[\alpha]_D$ +91 (water); Ref.: 138

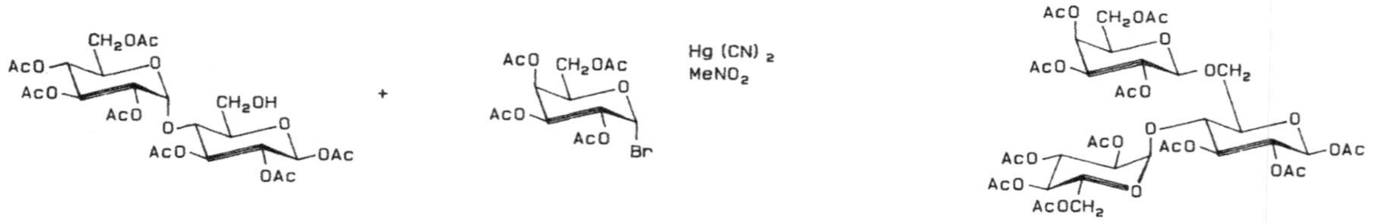

Hg(CN)₂
MeNO₂

52%; m.p. 175-176 °C; $[\alpha]_D$ +46.2 (chloroform); Ref.: 138

β-D-Glcp-(1→2) ⟍
                  ⟩ D-Glc
α-D-Glcp-(1→6) ⟋

+

AgOTf
DNBC1
Et₃N
dichloro-
methane

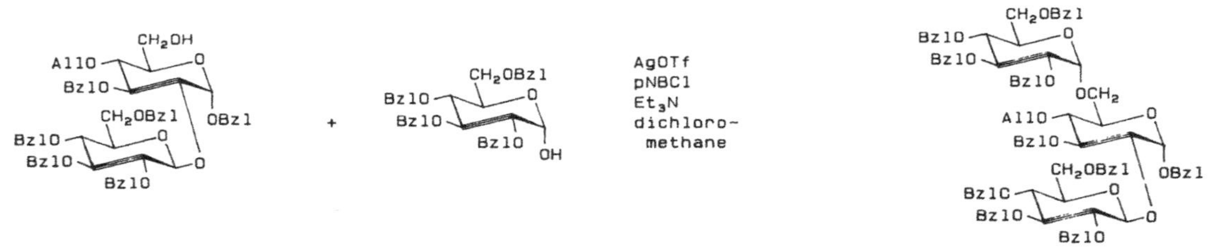

31%; [α]_D +69 (chloroform): C-13; Ref.: 139

β-D-Glcp-(1→3) ⟍
                  ⟩ D-Glc
α-D-Glcp-(1→6) ⟋

+

Me₃SiBr
CoBr₂
Et₄NBr
dichloro-
methane

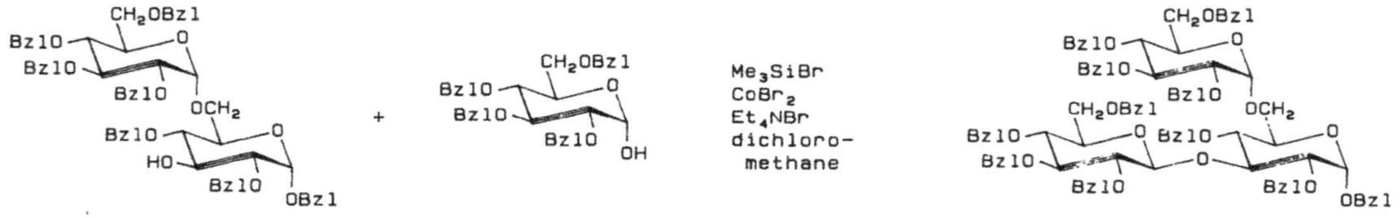

19%; [α]_D +66 (chloroform): C-13; Ref.: 130

β-D-Glcp-(1-2)
β-D-Glcp-(1-3) ⟩ D-Glc

β-D-(1-OMe): m.p. 249–251 °C; [α]$_D$ −32.0 (water); Ref.: 140

α-D-(1-OMe): amorphous powder; [α]$_D$ +34.8 (water); C-13: Ref.: 141

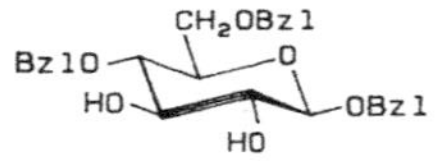

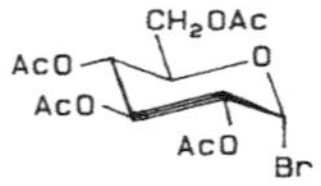

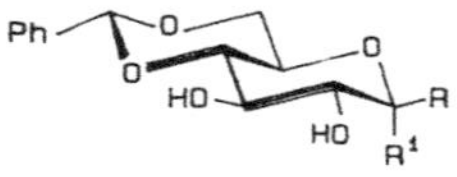

I
1 Ag$_2$O chloroform
2 PdC; H$_2$
3 NaOMe

II R=OMe; R^1=H
AgOTf; TMK tetrachloroethane

III R=H; R^1=OMe
1 Ag$_2$CO$_3$; I$_2$ dichloroethane
2 NaOMe
3 H$^+$

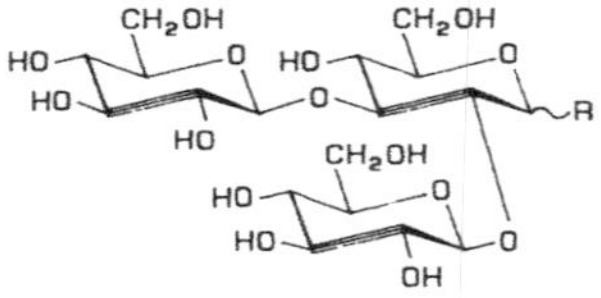

I R=OH; 1.5%; Ref.: 142

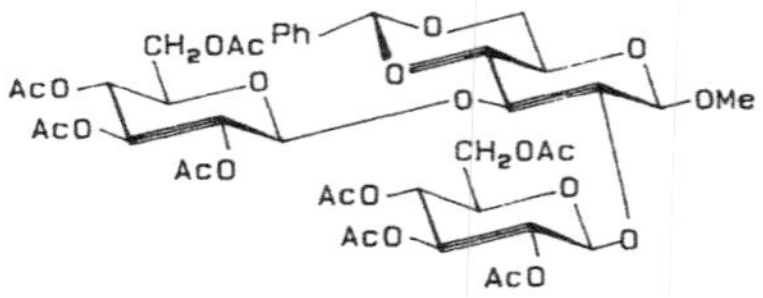

II 87%; [α]$_D$ −29.1 (chloroform); Ref.: 140

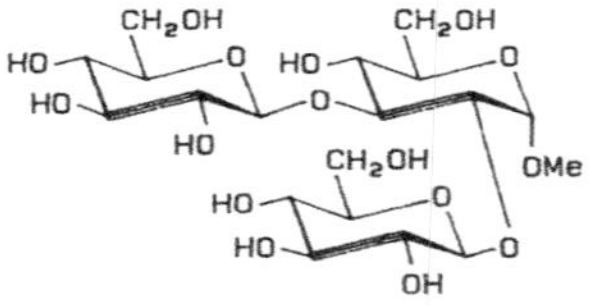

III 4.2%; Ref.: 141

β-D-Glcp-(1→2)
                    ⟩D-Glc
β-D-Glcp-(1→6)

α-ᴅ-(1→OMe): m.p. 262—265 °C; [α]ᴅ +28.4 (water); Ref.: 141

I   1 Ag₂CO₃; I₂
      dichloro-
      ethane
    2 H⁺
    3 NaOMe

II AgOTf
    pNBCl
    Et₃N
    dichloro-
    methane

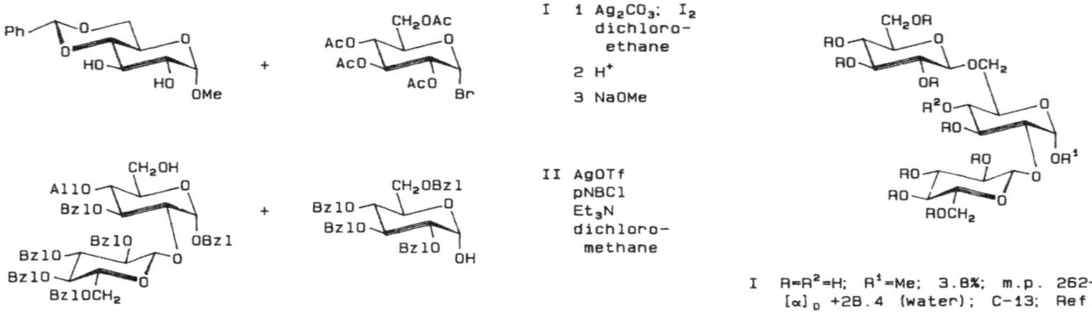

I   R=R²=H; R¹=Me; 3.8%; m.p. 262—265 °C;
    [α]ᴅ +28.4 (water); C-13; Ref.: 141

II R=R¹=Bzl; R²=All; 68%;
    [α]ᴅ +43 (chloroform); C-13; Ref.: 139

β-D-Glcp-(1-3)
β-D-Glcp-(1-6) >D-Glc

[α]_D +0.6 (water): Ref.: 143

m.p. 183-188 °C; [α]_D -1.1 (water): C-13; Ref.: 132

white powder: [α]_D -13.0 → +5.8 (water): C-13: Ref.: 144

α-p-(1-OMe): [α]_D +28.5 (MeOH): C-13: Ref.: 144

β-p-(1-OMe): m.p. 144-146 °C: [α]_D -38.4 (water): Ref.: 145

I   R¹=Bzl: Ref.: 143
    1 Ag₂O: I₂ chloroform
    2 NaOMe
    3 PdC: H₂

II R¹=Me
    Ag₂CO₃
    dichloroethane

III HgBr₂
    dichloroethane

IV AgOTf
    toluene
    MeNO₂

V   AgOTf
    pNBCl
    Et₃N
    dichloromethane

I   Ref.: 143

II R=Ac: R¹=H; R²=OMe; R³=H: 13%; m.p. 226-228 °C;
    [α]_D -30.5 (chloroform); Ref.: 145

III R=Ac: R¹=OMe; R²=H; R³=Bzl: 59.3%; Ref.: 144
    R=Ac: R¹=OBzl; R²=H; R³=Bzl: 47%;
    [α]_D +12.8 (chloroform): Ref.: 144

IV R=Bz: R¹=H; R²=OBzl; R³=Bzl: 77%;
    [α]_D +7 (chloroform); C-13: Ref.: 146

V   R=Bzl; R¹=OBzl; R²=H; R³=Bzl: 29%;
    [α]_D +38 (chloroform); C-13: Ref.: 132

β-D-GlcpA-(1→2) \
                  〉 D-Glc
β-D-Glcp-(1→4) /

[α]$_D$ -3.2 (water); Ref.: 56

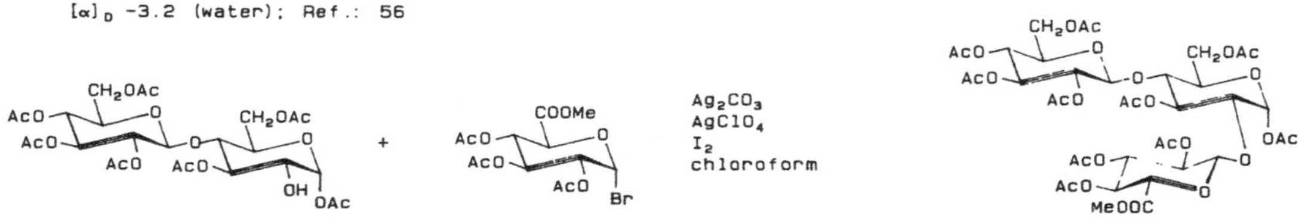

Ag$_2$CO$_3$
AgClO$_4$
I$_2$
chloroform

m.p. 272-276 °C; [α]$_D$ +11.4 (chloroform); Ref.: 56

β-D-Glcp-(1→4) \
                  〉 D-Glc
β-D-GlcpA-(1→6) /

[α]$_D$ +3.3 (water); Ref.: 56

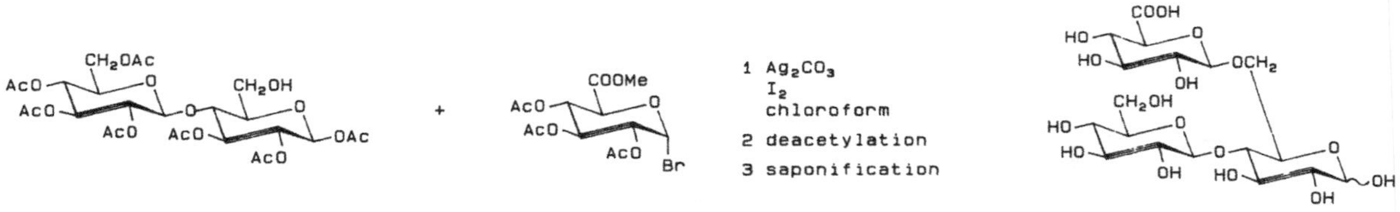

1 Ag$_2$CO$_3$
  I$_2$
  chloroform
2 deacetylation
3 saponification

34%; Ref.: 56

α-D-Manp-(1-2)
α-L-Rhap-(1-6) ⟩ D-Glc

hygroscopic white amorphous powder; m.p. 180-182°C (dec.); [α]$_D$ +40.6 (water); Ref.: 147

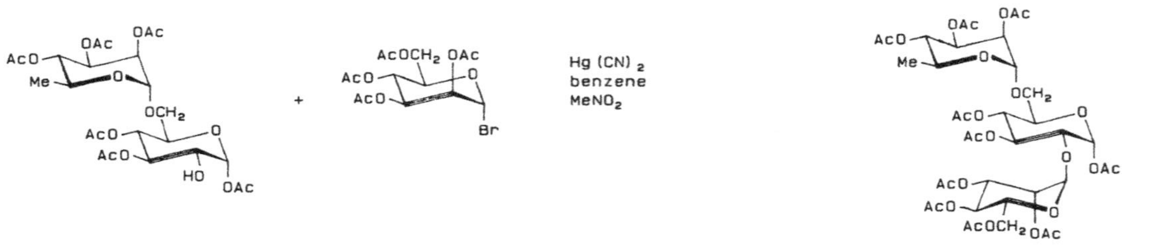

Hg(CN)$_2$
benzene
MeNO$_2$

+

65%; m.p. 87-88°C; [α]$_D$ +12.5 (methanol); Ref.: 147

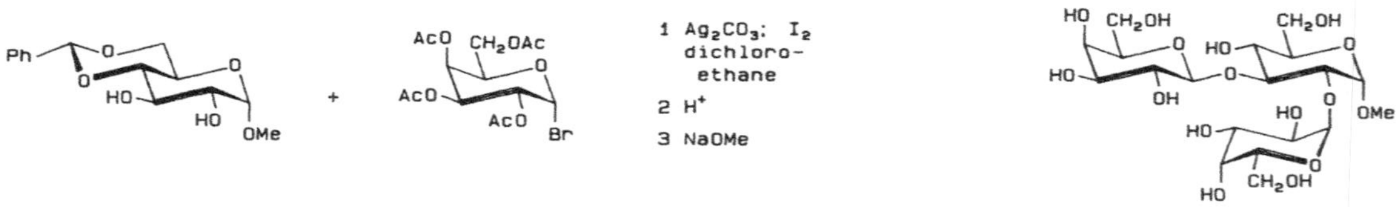

1.4%; Ref.: 148

α-D-Galp-(1-2)
β-D-Galp-(1-3) ⟩ D-Glc

α-D-(1-OMe); [α]$_D$ +107 (water); Ref.: 148

1 Ag$_2$CO$_3$; I$_2$
  dichloro-
  ethane
2 H$^+$
3 NaOMe

+

α-D-Galp-(1→3)
⟩D-Glc
α-D-Galp-(1→6)

α-p-(1-OMCO): [α]_D +79.2 (water); C-13; Ref.: 149

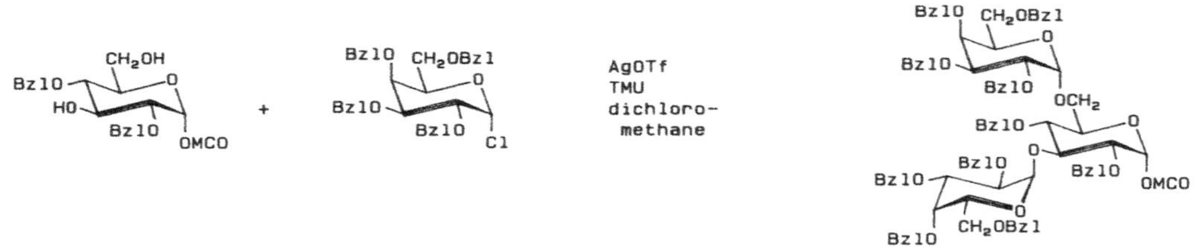

BzlO CH₂OBzl
BzlO—O
BzlO—Cl

+

CH₂OH
BzlO—O
HO—BzlO—OMCO

AgOTf
TMU
dichloro-
methane

57%; [α]_D +53.1 (chloroform); C-13; Ref.: 149

BzlO CH₂OBzl
BzlO—O
BzlO OCH₂
BzlO—O
BzlO—OMCO

BzlO—O
BzlO—O
BzlO—CH₂OBzl

β-D-Galp-(1→2)
⟩D-Glc
β-D-Galp-(1→3)

α-p-(1-OMe): m.p. 247°C; [α]_D +60 (water); Ref.: 148

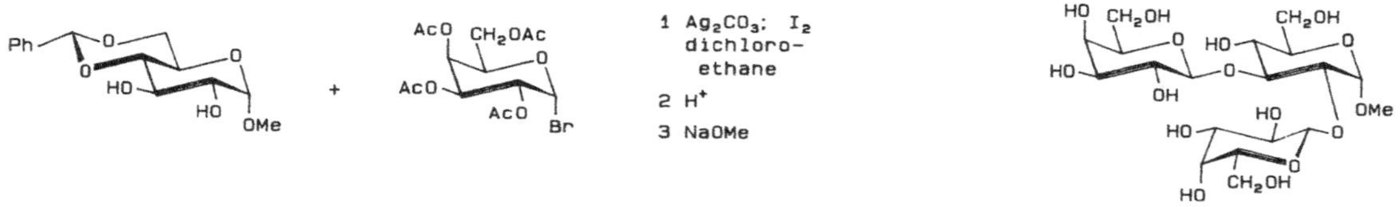

AcO CH₂OAc
AcO—O
AcO—Br

+

Ph—O—O
HO—O—OMe
HO

1 Ag₂CO₃; I₂
  dichloro-
  ethane
2 H⁺
3 NaOMe

CH₂OH
HO—O
HO—O—OMe

HO CH₂OH
HO—O
OH

HO—O
HO—CH₂OH

3.7%; Ref.: 148

β-D-Galp-(1→3)
⟩D-Glc
β-D-Galp-(1→6)

α-D-(1→OMe): m.p. 244—246 °C; [α]_D +60 (water): Ref.: 148

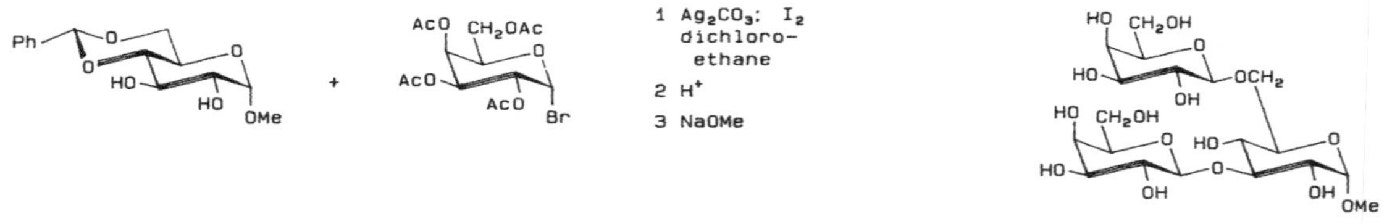

1 Ag₂CO₃; I₂
   dichloro-
   ethane
2 H⁺
3 NaOMe

4.1%; Ref.: 148

α-L-Fucp-(1→3)
⟩D-Glc
β-D-Galp-(1→4)

[α]_D −43 (methanol); C-13; Ref.: 113

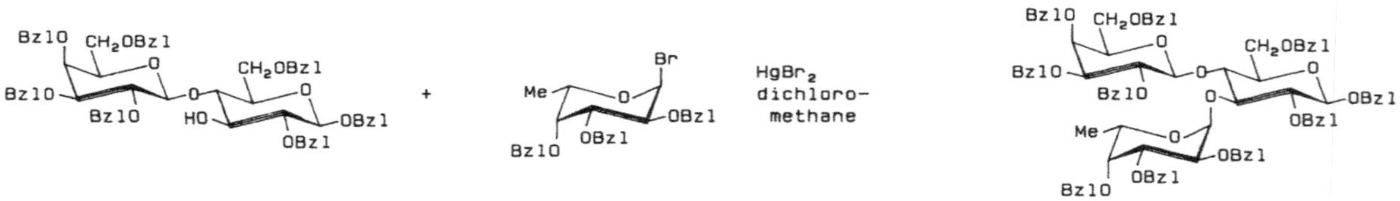

HgBr₂
dichloro-
methane

53%; [α]_D −43 (chloroform); C-13; Ref.: 113

α-Colp-(1→3)
                    \
                     D-Glc
α-Colp-(1→6)   /

α-D-(1→OMCO): [α]_D −17.7 (water); C-13; Ref.: 149

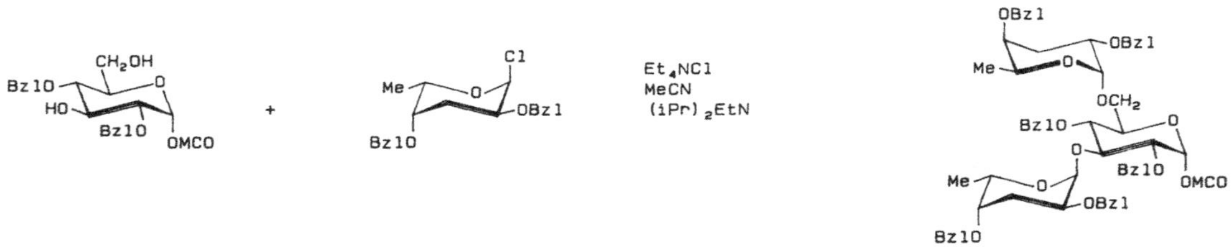

Et_4NCl
MeCN
(iPr)_2EtN

53%; [α]_D −8.5 (chloroform); C-13; Ref.: 149

Colp = 3,6-Dideoxy-α-L-Xylo-Hexp

β-D-Glcp-(1→4)-β-D-Glcp-(1→5)-D-GlcA

Ag_2O; I_2
chloroform

42%; m.p. 153−154 °C; [α]_D −4.6 (chloroform); Ref.: 150

α, β-D-Glcp-(1→3)
α, β-D-Glcp-(1→4) D-GlcA

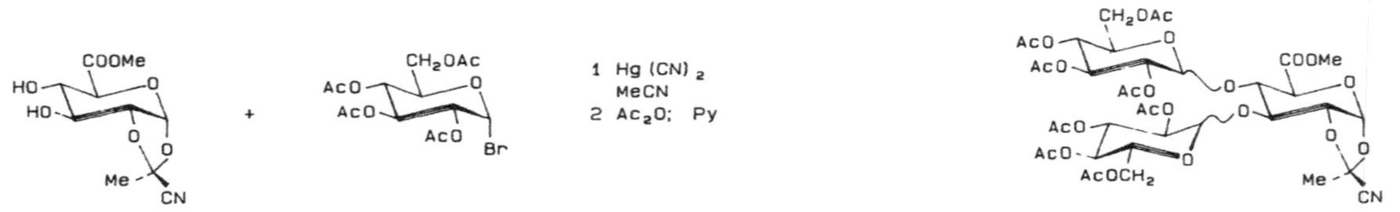

CH₂OAc
AcO
AcO   AcO   Br

+

COOMe
HO
HO   Me   CN

1 Hg(CN)₂
  MeCN
2 Ac₂O; Py

CH₂OAc
AcO
AcO   AcO

COOMe
Me   CN

CH₂OAc
AcO
AcO   AcO   AcOCH₂

4%; [α]_D +21.1 (chloroform); Ref.: 151

α-L-Rhap-(1→3)
α-L-Rhap-(1→4) D-GlcA

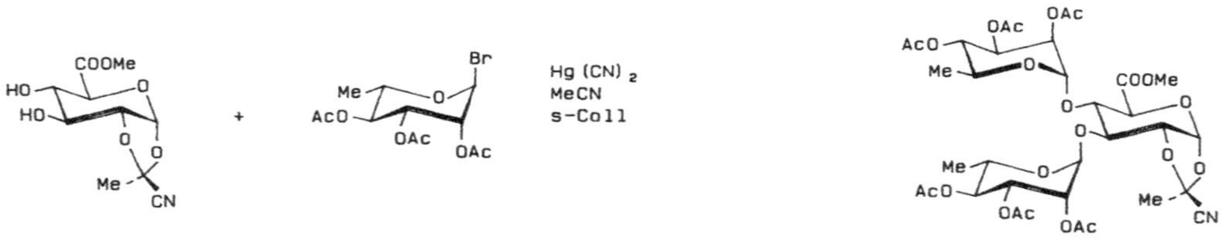

Br
Me
AcO   OAc   OAc

Hg(CN)₂
MeCN
s-Coll

+

COOMe
HO
HO   Me   CN

OAc
OAc
AcO   Me

COOMe
Me   CN

Me
AcO   OAc   OAc

14%; [α]_D -39.6 (chloroform); C-13; Ref.: 151

α-D-GlcpN-(1→6)-α-D-GlcpN-(1→6)-D-GlcN

.3HCl; [α]$_D$ +92 (water): Ref.: 152

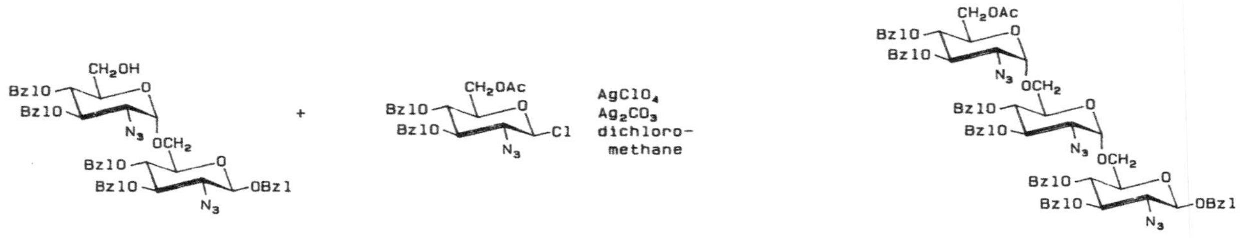

AgClO$_4$
Ag$_2$CO$_3$
dichloro-
methane

m.p. 80-82 °C; [α]$_D$ +73 (chloroform): Ref.: 152

α-D-GlcpNSO$_3$(3,6-OSO$_3$)-(1→4)-α-L-IdopA(2-OSO$_3$)-(1→4)-D-GlcNSO$_3$(6-OSO$_3$)

sodium salt; colourless glass; [α]$_D$ +35 (water): Ref.: 153

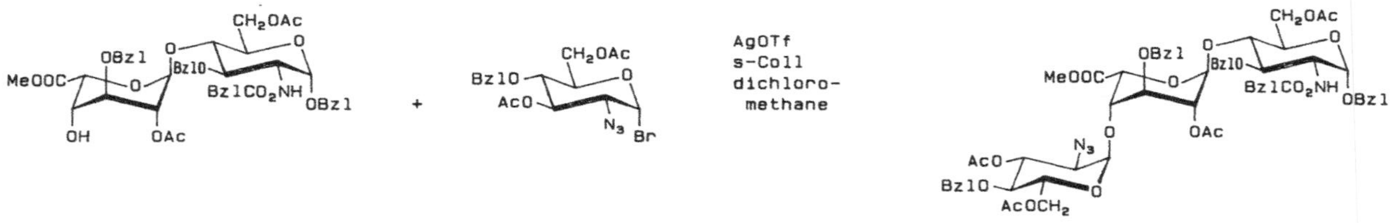

AgOTf
s-Coll
dichloro-
methane

88%; [α]$_D$ +83 (chloroform): Ref.: 153

α-KDO-(2→4)-α-KDO-(2→6)-D-GlcN

[α]$_D$ +82 (water); Ref.: 154

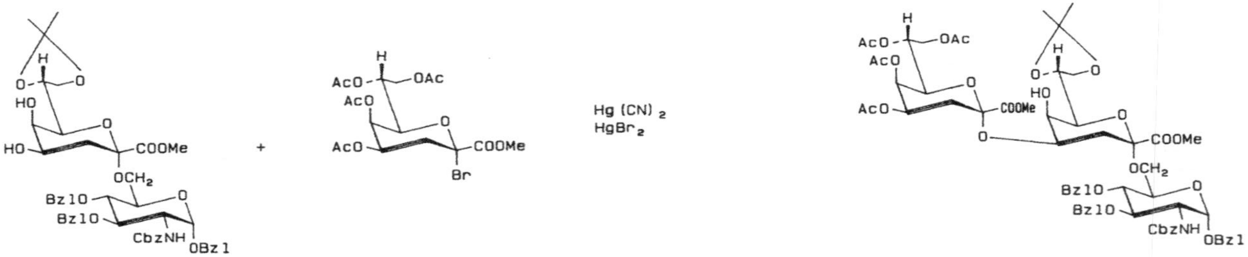

Hg(CN)$_2$
HgBr$_2$

44%; Ref.: 154

β-D-Glcp-(1→4)-β-D-Glcp-(1→4)-D-GlcNAc

[α]$_D$ +30 (water); Ref.: 155

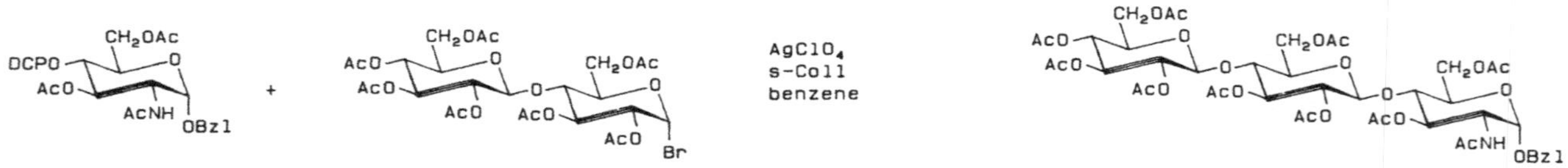

AgClO$_4$
s-Coll
benzene

36%; m.p. 179—180 °C; [α]$_D$ +35 (chloroform); Ref.: 155

(DCP=2,3-diphenyl-2-cyclopropenyl)

β-D-Galp-(1→4)-β-D-Glcp-(1→4)-D-GlcNAc

m.p. 200—203 °C; [α]_D +24 → +15 (water); Ref.: 156

α-p-(1-OBzl); m.p. 310—311°C; [α]_D +76 (water); Ref.: 156

AgClO_4
DCP-ClO_4
s-Coll
benzene

42%; [α]_D +41 (chloroform); Ref.: 156

α-L-Fucp-(1→2)-β-D-Glcp-(1→4)-D-GlcNAc

β-OMCO; [α]_D -78.7 (water); C-13; Ref.: 157

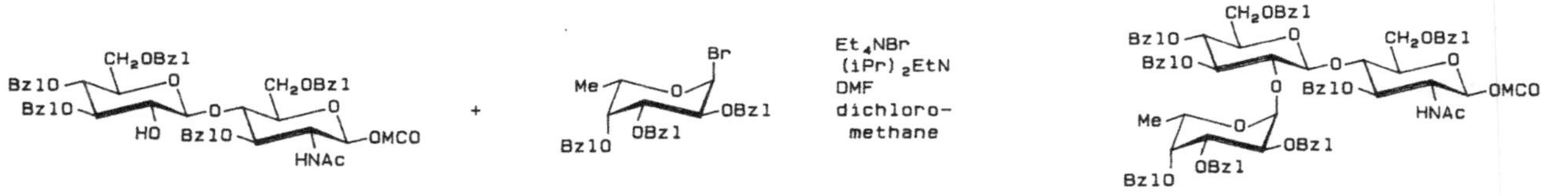

Et_4NBr
(iPr)_2EtN
DMF
dichloro-
methane

57%; [α]_D -13.0 (chloroform); C-13; Ref.: 157

β-D-Galp-(1→4)-β-D-Glcp-(1→6)-D-GlcNAc

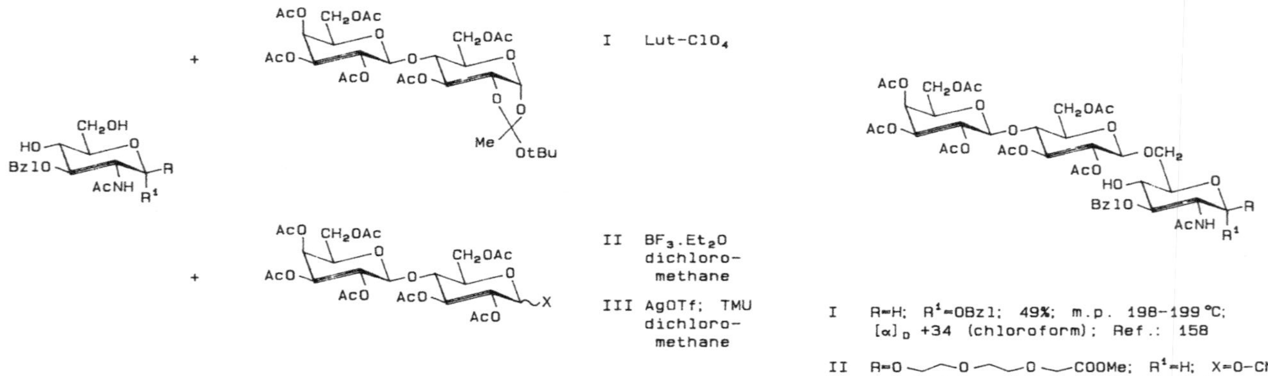

I   Lut–ClO₄

II BF₃·Et₂O
    dichloro-
    methane

III AgOTf; TMU
    dichloro-
    methane

I   R=H; R¹=OBzl; 49%; m.p. 198-199°C;
    [α]_D +34 (chloroform); Ref.: 158

II R=O⌒O⌒O—COOMe; R¹=H; X=O-CNH-CCl₃; 51.5%;
    [α]_D -23 (chloroform); Ref.: 159

III R=O⌒O⌒O—N₃; R¹=H; X=α-Br; 41%;
    [α]_D -16 (chloroform); Ref.: 159

α-D-GlcpN-(1→4)-β-D-GlcpA-(1→4)-D-GlcN

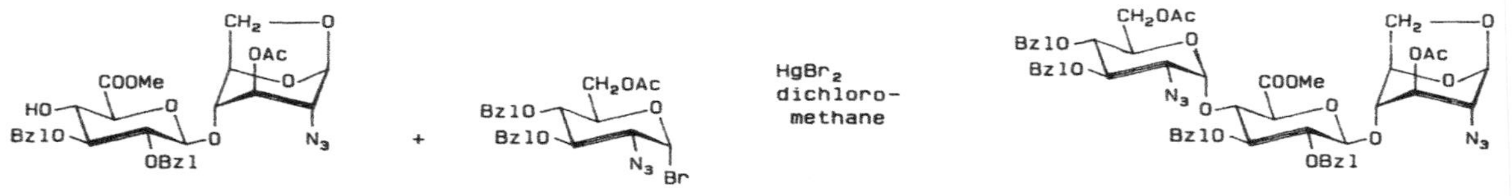

HgBr₂
dichloro-
methane

69%; [α]_D +28 (chloroform); Ref.: 160

α-GlcpN-(1→3)-α-D-GlcpN-(1→3)-D-GlcN

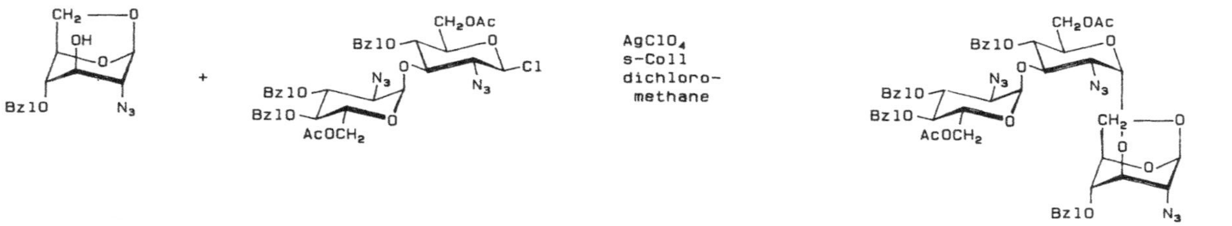

AgClO₄
s-Coll
dichloro-
methane

62%; [α]_D +70 (chloroform); Ref.: 161

α-D-GlcpN-(1-6)-α-D-GlcpN-(1-6)-D-GlcN

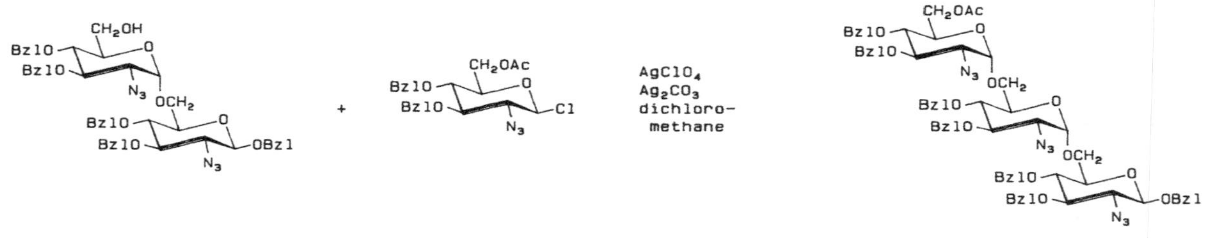

71%; m.p. 80-82°C; [α]$_D$ +72 (chloroform); Ref.: 152, 162

β-D-GlcpN-(1-6)-α-D-GlcpN-(1-6)-D-GlcN

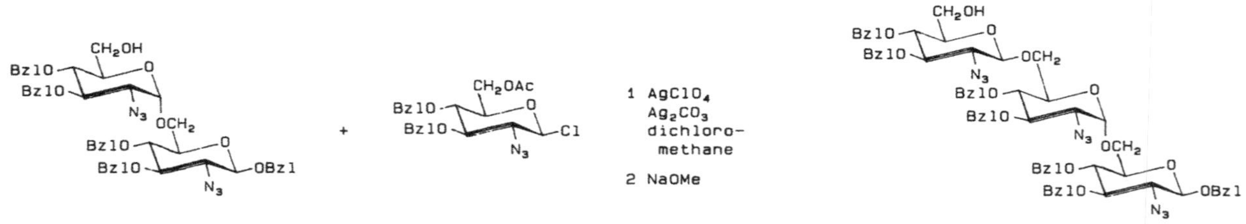

18%; [α]$_D$ +21 (chloroform); Ref.: 152

β-D-GlcpN-(1→3)-β-D-GlcpN-(1→3)-D-GlcN

β-D-GlcpNAc-(1→3)-β-D-GlcpNAc-(1→3)-D-GlcNAc

[α]$_{578}$ +15.4 (water); Ref.: 163

β-acetate: m.p. 194°C; [α]$_{578}$ +7.0 (chloroform); C-13; Ref.: 163

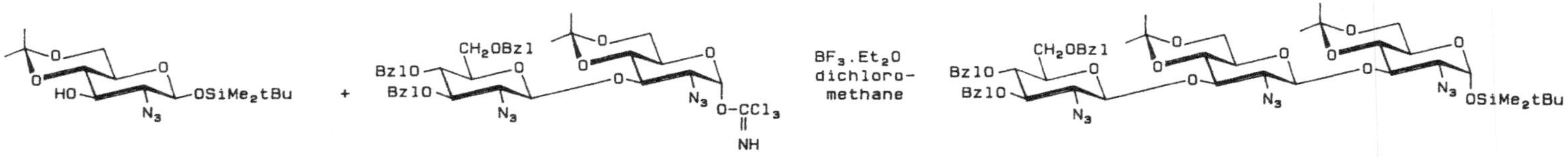

BF$_3$·Et$_2$O
dichloro-
methane

70%; m.p. 106—108°C; [α]$_{578}$ −34.7 (chloroform); C-13; Ref.: 163

β-D-Glcp-(1→4)-β-D-GlcpNAc-(1→4)-D-GlcNAc

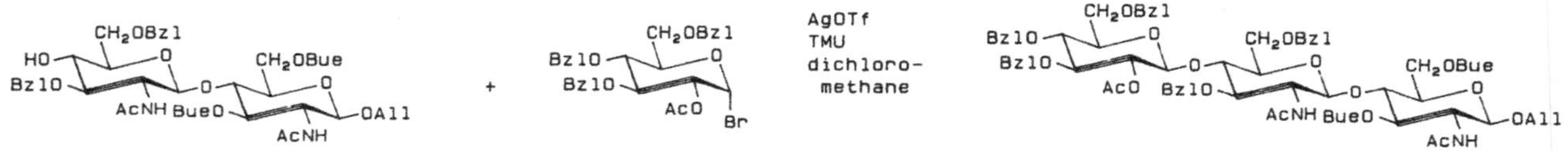

AgOTf
TMU
dichloro-
methane

44%; [α]$_D$ −26.5 (chloroform); Ref.: 164

β-D-GlcpNAc-(1→6)-β-D-GlcpNAc-(1→4)-D-GlcNAc

β-D-(1-O-p-Nitrophenyl); m.p. 257.5-259 °C; [α]$_D$ -2 (water); Ref.: 165

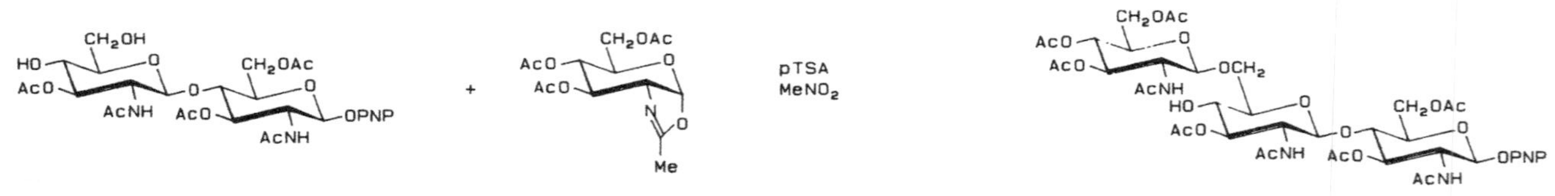

pTSA
MeNO$_2$

14%; m.p. 273.5-275 °C; [α]$_D$ -44 (DMF); Ref.: 165

α-D-Manp-(1→4)-β-D-GlcpN-(1→4)-D-GlcN

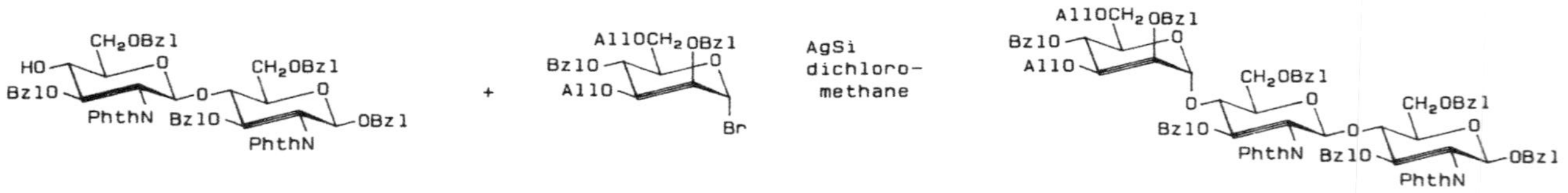

AgSi
dichloro-
methane

36%; [α]$_D$ +15.6 (chloroform); Ref.: 166

α-D-Manp-(1→6)-α-D-GlcpNAc-(1→4)-D-GlcNAc

m.p. 186-188 °C; [α]$_D$ +36 → +34 (methanol): Ref.: 167

β-NHCOCH$_2$CH(NH$_2$)-COOH; m.p. 160 °C (dec. without melting): [α]$_D$ +39 (methanol): Ref.: 168

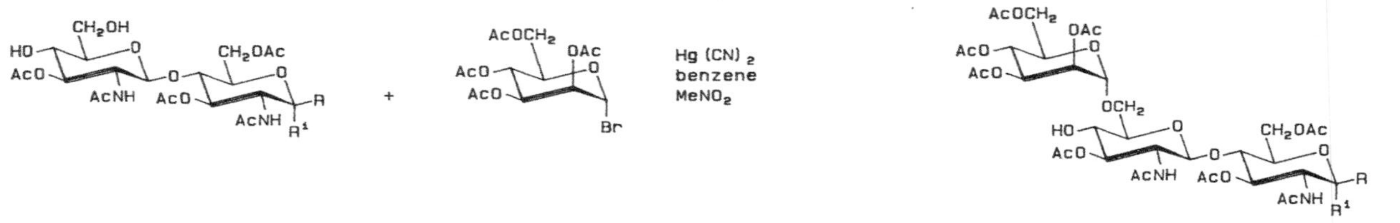

R=H; R^1=OAc; 66%; m.p. 136-138 °C;
[α]$_D$ +29 (chloroform): Ref.: 167

R=N$_3$; R^1=H; 48%; m.p. 133-134 °C:
[α]$_D$ -12.3 (chloroform): Ref.: 168

Hg(CN)$_2$
benzene
MeNO$_2$

β-D-Manp-(1→4)-β-D-GlcpN-(1→4)-D-GlcN

β-D-Manp-(1→4)-β-D-GlcpNAc-(1→4)-D-GlcNAc

amorphous solid: [α]$_D$ +0.5 (water): Ref.: 164
amorphous solid: [α]$_D$ +0.2 (water): Ref.: 169
amorphous powder: [α]$_D$ +0.2 (water): Ref.: 170

I   pTSA dichloro-ethane

II AgOTf s-Coll dichloro-methane

III 1 AgOTf TMU dichloro-methane
    2 NaOMe
    3 Ac₂O; DMSO
    4 NaBH₄

IV AgSi dichloro-methane

I   R²=OBzl: 25%; [α]$_D$ +23 (chloroform): Ref.: 169

II R¹=OBzl; R²=H: 50%; m.p. 52 °C; [α]$_D$ +33.6 (chloroform): Ref.: 170

III R²=OAll: m.p. 165.5-169 °C; [α]$_D$ -31.6 (chloroform): Ref.: 164

IV R¹=H; R²=OBzl: 40%; [α]$_D$ -2.1 (chloroform): Ref.: 166

β-D-GlcpNAc-(1→3)-β-D-GlcpNAc-(1→6)-D-GlcNAc

β-p-(1-O-p-Nitrophenyl); m.p. 209—210 °C; [α]$_D$ +8 (water-MeOH); Ref.: 171

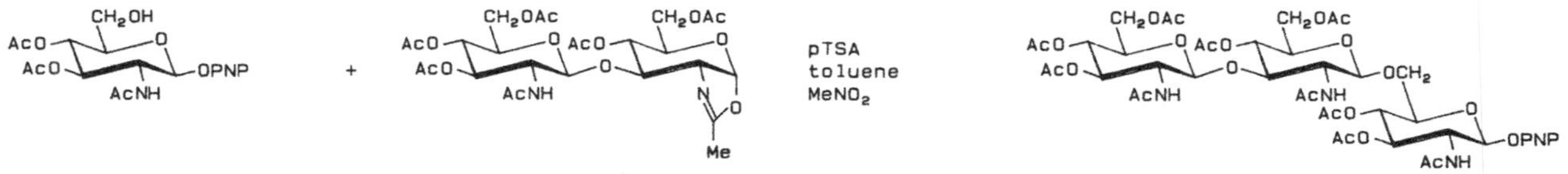

pTSA
toluene
MeNO$_2$

78%; m.p. 280—281 °C; [α]$_D$ +75 (chloroform-MeOH); Ref.: 171

β-D-GlcpNAc-(1→4)-β-D-GlcpNAc-(1→6)-D-GlcNAc

β-p-(1-O-Nitrophenyl); m.p. 239—241 °C; [α]$_D$ -47.5 (water); Ref.: 172

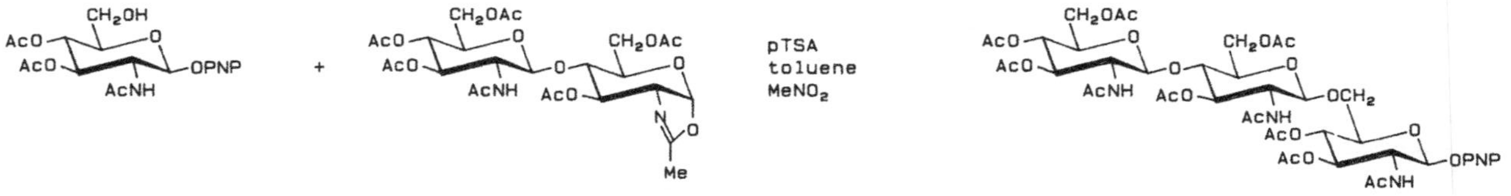

pTSA
toluene
MeNO$_2$

30%; m.p. 289—290 °C; [α]$_D$ -31 (MeOH-MeNO$_2$ 1:1); Ref.: 172

β-D-Manp-(1→4)-β-D-GlcpN-(1→6)-D-GlcN

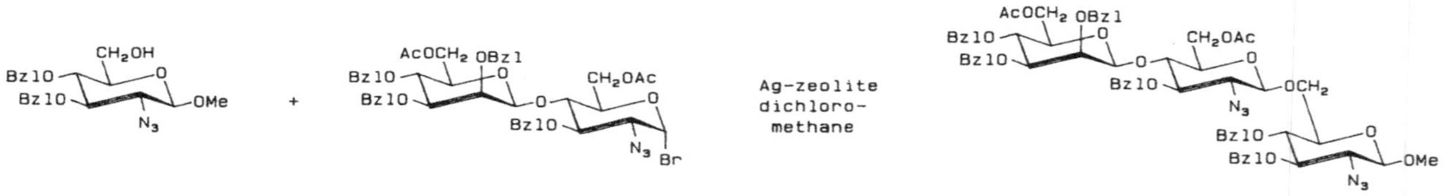

60%; [α]$_D$ -25 (chloroform); Ref.: 173

Ag-zeolite
dichloro-
methane

α-KDO-(2→3)-β-D-GlcpN-(1→6)-D-GlcN
α-KDO-(2→3)-β-D-GlcpNAc-(1→6)-D-GlcNAc

amorphous; [α]$_D$ +16 (water); Ref.: 174

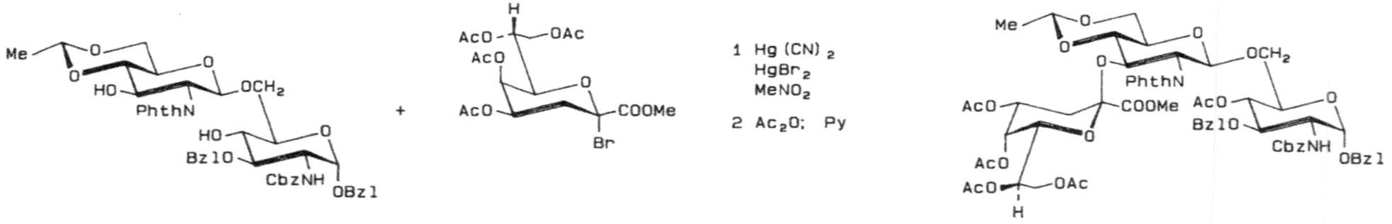

1 Hg(CN)$_2$
  HgBr$_2$
  MeNO$_2$

2 Ac$_2$O; Py

13%; m.p. 121°C; [α]$_D$ +58 (chloroform); Ref.: 174

α-KDO-(2→6)-β-D-GlcpN-(1→6)-D-GlcN

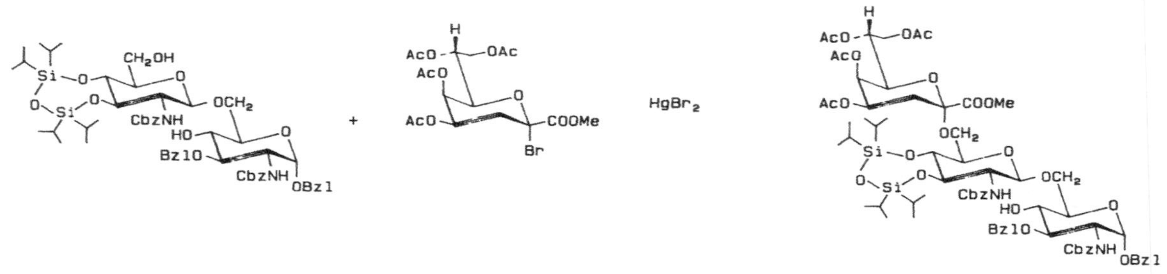

HgBr$_2$

40%; Ref.: 154

β-KDO-(2→3)-β-D-GlcpNAc-(1→6)-D-GlcN

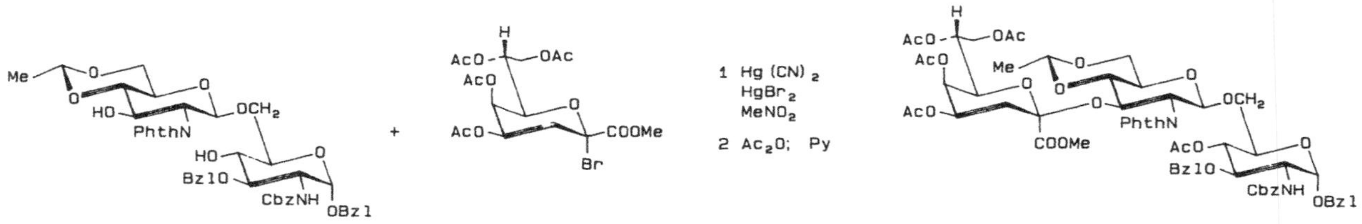

1 Hg(CN)$_2$
   HgBr$_2$
   MeNO$_2$
2 Ac$_2$O; Py

1.3%; [α]$_D$ +40 (chloroform); Ref.: 174

β-D-GlcpN-(1→4)-β-D-Manp-(1→4)-D-GlcN

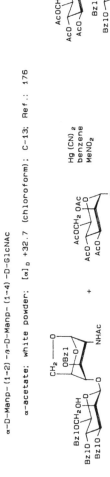

78%; [α]_D +15.5 (chloroform); Ref.: 175

α-D-Manp-(1→2)-β-D-Manp-(1→4)-D-GlcNAc

α-acetate: white powder; [α]_D +32.7 (chloroform); C-13; Ref.: 176

40.7%; [α]_D -29 (chloroform); C-13; Ref.: 176

α-D-Manp-(1→3)-β-D-Manp-(1→4)-D-GlcN

α-D-Manp-(1→3)-β-D-Manp-(1→4)-D-GlcNAc

syrup: $[\alpha]_D$ +27.8 (methanol): Ref.: 177

α-acetate: $[\alpha]_D$ +3.9 (chloroform): Ref.: 177

α,β-acetate: white powder: $[\alpha]_D$ +20 (chloroform): C-13: Ref.: 176

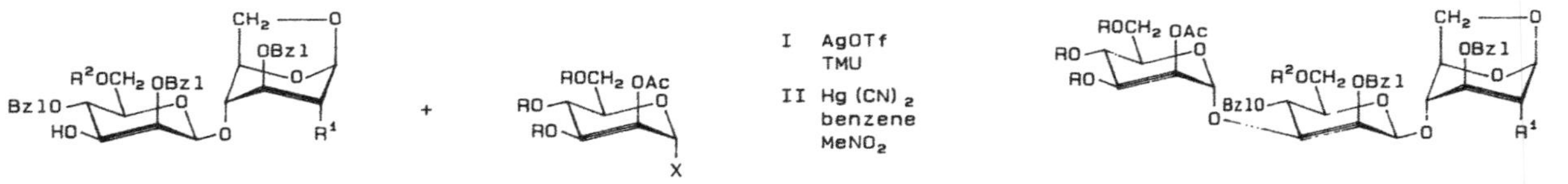

I   AgOTf
    TMU

II Hg(CN)$_2$
    benzene
    MeNO$_2$

I   R=Bzl; R^1=N$_3$; R^2=Ac; X=Cl: 80%:
    $[\alpha]_D$ -0.5 (chloroform): C-13: Ref.: 177

II R=Ac; R^1=NHAc; R^2=Bzl; X=Br; 47.1%:
    $[\alpha]_D$ -27.8 (chloroform): C-13: Ref.: 176

α-D-Manp-(1→6)-α-D-Manp-(1→4)-D-GlcN

α-D-Manp-(1→6)-α-D-Manp-(1→4)-D-GlcNAc

syrup: $[\alpha]_D$ +36.5 (methanol); Ref.: 177

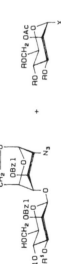

AgOTf
TMU
dichloro-
methane

R=Ac; R¹=Bzl; X=Br; 90%; $[\alpha]_D$ +8.7 (chloroform); Ref.: 177

R=Ac; R¹=All; X=Br; 91%; $[\alpha]_D$ +16.6 (chloroform); Ref.: 177

R=Bzl; R¹=All; X=Cl; 90%; $[\alpha]_D$ +5.1 (chloroform); Ref.: 177

α-L-Rhap-(1→3)-α-L-Rhap-(1→3)-D-GlcNAc

β-OMCO: [α]$_{589}$ -55 (methanol): C-13; Ref.: 178

I AgOTf
   TMU
   dichloro-
   methane

I AgOTf
   TMU
   dichloro-
   methane

II Hg(CN)$_2$
    MeCN
    MeNO$_2$

III Hg(CN)$_2$
     benzene
     MeCN

I   R=Bz; R^1=H; R^2=OMCO; R^3=Bzl; 67%; m.p. 153-155 °C;
     [α]$_{589}$ +5 (chloroform): C-13; Ref.: 178

     R=Bz; R^1=H; R^2=OMCO; R^3=Ac; 61%; m.p. 108-110 °C;
     [α]$_{589}$ -9.6 (chloroform): C-13; Ref.: 178

II   R=H; R^1=H; R^2=OMCO; R^3=Ac; 25%; m.p. 123-124 °C;
     [α]$_D$ -80 (methanol); Ref.: 179

III R=H; R^1=OMCO; R^2=H; R^3=Ac; 63%; m.p. 102 °C;
     [α]$_D$ -16 (methanol); Ref.: 179

α-L-Rhap-(1→4)-α-L-Rhap-(1→3)-D-GlcNAc

10%; m.p. 119°C; [α]$_D$ -92 (methanol): Ref.: 179

α-L-Fucp-(1→2)-α-D-Galp-(1→3)-D-GlcNAc

β-D-(1-O-p-Nitrophenyl): m.p. 262-264°C; [α]$_D$ +14.3 (DMSO): Ref.: 180

44%; [α]$_D$ -4.7 (MeOH); C-13: Ref.: 180

α-D-Galp-(1-4)-β-D-Galp-(1-3)-D-GlcNAc

β-p-(1-OPr); amorphous solid; $[\alpha]_D$ +43.5 (MeOH); C-13; Ref.: 181

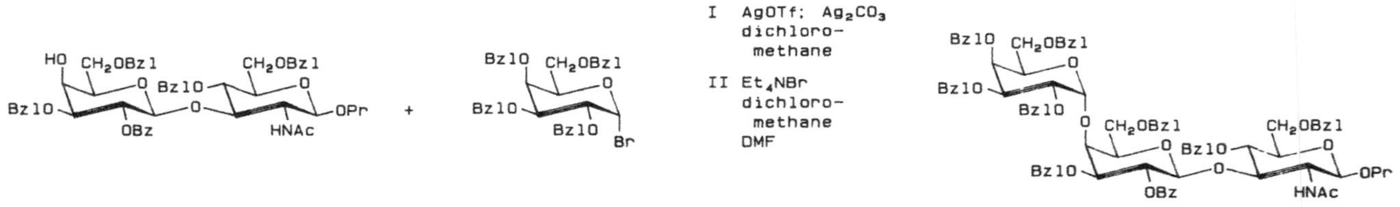
HO
CH₂OBzl
BzlO
OBz
HNAc
OPr
+
BzlO
Br
I AgOTf; Ag₂CO₃
dichloro-
methane
II Et₄NBr
dichloro-
methane
DMF

I 72%; $[\alpha]_D$ +50.9; $[\alpha]_{436}$ +104 (chloroform); Ref.: 181

II 34%; Ref.: 181

α-D-GalpN-(1→3)-β-D-Galp-(1→3)-D-GlcN

α-D-GalpNAc-(1→3)-β-D-Galp-(1→3)-D-GlcNAc

$[\alpha]_D$ +135.1 (water): Ref.: 182

amorphous; $[\alpha]_D$ +130 (water): Ref.: 183

$[\alpha]_D$ +130 (water): Ref.: 186

I Ag_2CO_3; $AgClO_4$ dichloromethane

I R^1=H

II 1 $Hg(CN)_2$; $HgBr_2$ benzene, $MeNO_2$
    2 deacetylation
    R^1=Ac

I $R=R^1=R^2$=Ac; 45%; m.p. 87 °C;
    $[\alpha]_D$ +76.8 (chloroform); Ref.: 182, 184

    R, R=PhCH; R^1=H; R^2=Bzl; 33%; m.p. 265 °C;
    $[\alpha]_D$ +91 (chloroform); Ref.: 183, 185, 186

II R, R=PhCH; R^1=H; R^2=Bzl; 74%; Ref.: 183, 186

α-L-Fucp-(1-2)-β-D-Galp-(1-3)-D-GlcNAc

[α]$_D$ -15.4; Ref.: 187

[α]$_D$ -30 (water); Ref.: 188

amorphous; [α]$_D$ -19.4 (MeOH); Ref.: 189

[α]$_D$ -18 (MeOH); Ref.: 183

[α]$_D$ -18 (MeOH); Ref.: 186

β-p-(1-O-p-Nitrophenyl); amorphous; [α]$_D$ -52.5 (DMSO); C-13; Ref.: 180

I 11%; [α]$_D$ -42.6 (MeOH); C-13; Ref.: 180

II R=R^1=Bzl; R^2=NBz; 25%;
[α]$_D$ -55.2 (chloroform); Ref.: 188

III R, R=PhCH; R^1=H; R^2=NBz; 65%; m.p. 166—169 °C
[α]$_D$ -43 (chloroform); Ref.: 183, 186

IV R, R=PhCH; R^1=Bz; R^2=Bzl; 82%; m.p. 114 °C
[α]$_D$ +69.8 (chloroform); Ref.: 187, 189

I 1 Hg(CN)$_2$
benzene
MeNO$_2$
2 H$^+$

II Hg(CN)
benzene
MeNO$_2$

III 1 Hg(CN)$_2$; HgBr$_2$
benzene
MeNO$_2$
2 thiocarbamide

IV Et$_4$NBr
dichloro-
methane
DMF

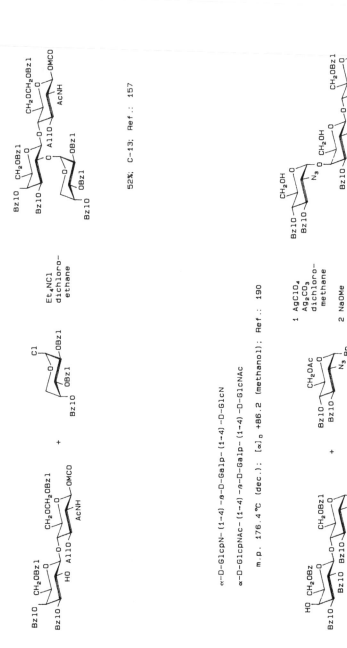

β-D-GlcpN-(1→6)-β-D-Galp-(1→4)-D-GlcN

β-D-GlcpNAc-(1→6)-β-D-Galp-(1→4)-D-GlcNAc

β-D-(1-OMe): C-13; Ref.: 191

AgOTf
s-Coll
MeNO$_2$

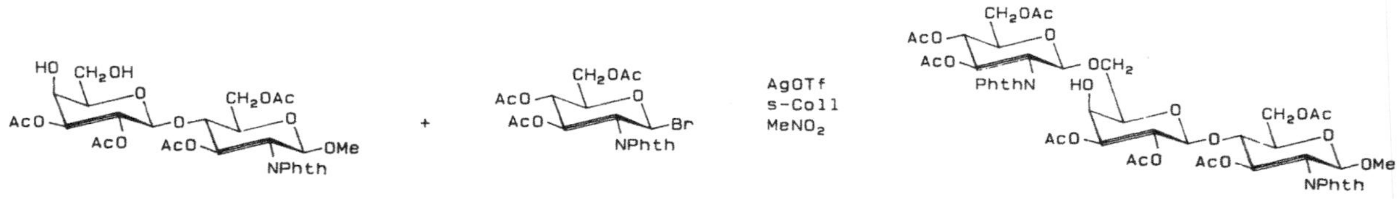

α-D-Galp-(1→3)-β-D-Galp-(1→4)-D-GlcNAc

m.p. 236-239°C; (dec.); $[\alpha]_D$ +104 → +100 (water): Ref.: 192

β-OMCO: $[\alpha]_D$ +49 (water); C-13; Ref.: 193

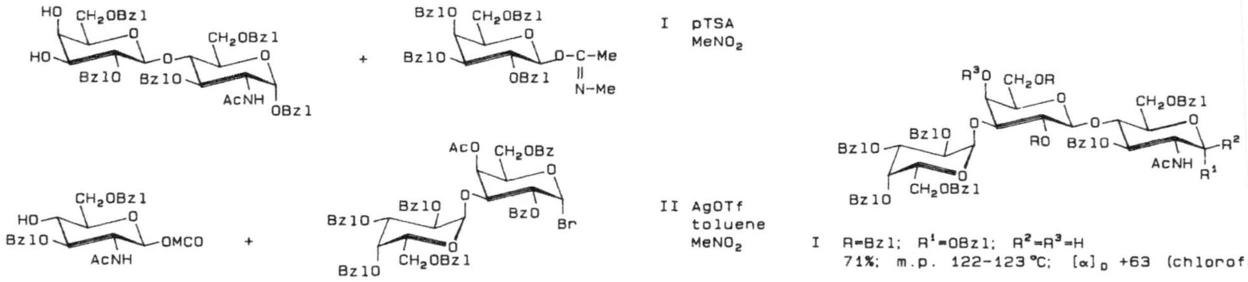

I   pTSA
    MeNO$_2$

II AgOTf
    toluene
    MeNO$_2$

I   R=Bzl; R^1=OBzl; R^2=R^3=H
    71%; m.p. 122-123°C; $[\alpha]_D$ +63 (chloroform): Ref.: 192

II R=Bz; R^1=H; R^2=OMCO; R^3=Ac
    73%; $[\alpha]_D$ +30 (chloroform); C-13; Ref.: 193

α-D-Galp-(1→4)-β-D-Galp-(1→4)-D-GlcN

α-D-Galp-(1→4)-β-D-Galp-(1→4)-D-GlcNAc

m.p. 179-180 °C; $[\alpha]_D$ +68 → +72 (MeOH-water, 9:1): Ref.: 194

β-D-(1-OPr): amorphous solid; $[\alpha]_D$ +18.8 (MeOH): C-13; Ref.: 195

β-D-(1-OPr): $[\alpha]_D$ +32 (MeOH-water, 9:1): Ref.: 194

β-D-(1-O—S—COOMe): $[\alpha]_D$ +34 (water): C-13; Ref.: 196

β-D-(1-O-2-(Octadecylthio)ethyl): $[\alpha]_D$ +35 (DMSO): C-13; Ref.: 196

β-D-(1-OEt): $[\alpha]_D$ +38 (water): C-13; Ref.: 196

I   AgOTf
     TMU
     dichloro-
     methane

II   AgOTf
     Ag2CO3
     dichloro-
     methane

III AgOTf
     s-Coll
     dichloro-
     methane

I   41%; m.p. 98-101 °C; $[\alpha]_D$ +65 (chloroform); C-13; Ref.: 196

II R^1=H; R^2=OAll; R^3=Bz; R^4=R^5=Bzl
     64%; $[\alpha]_D$ +38.6 (chloroform); Ref.: 195

III R^1=OBzl; R^2=H; R^3=R^4=R^5=Bzl
     82%; $[\alpha]_D$ +59 (chloroform); C-13; Ref.: 194

     R^1=H; R^2=OAll; R^3=R^4=Bzl; R^5=Bz
     78%; $[\alpha]_D$ +27 (chloroform); Ref.: 194

α-D-GalpN-(1→3)-α-D-Galp-(1→4)-D-GlcN

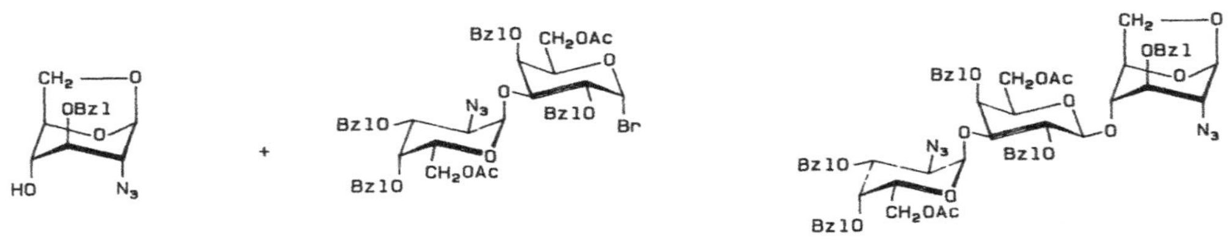

Ref.: 184

α-L-Fucp-(1→2)-β-D-Galp-(1→4)-D-GlcN

α-L-Fucp-(1→2)-β-D-Galp-(1→4)-D-GlcNAc

[α]$_D$ -46.5 (water): Ref.: 197

[α]$_D$ -46.4 (water): Ref.: 198

[α]$_D$ -55.9 (water): Ref.: 199

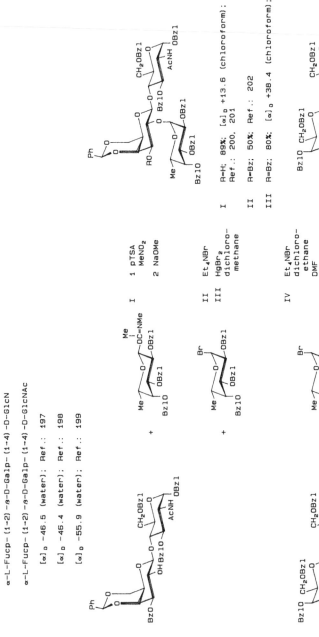

I   1 pTSA
      MeNO$_2$
    2 NaOMe

II   Et$_4$NBr

III HgBr$_2$
     dichloro-
     methane

IV Et$_4$NBr
     dichloro-
     ethane
     DMF

V   Et$_4$NBr
     (iPr)$_2$EtN
     dichloro-
     methane
     DMF

I   R=H; 89%; [α]$_D$ +13.6 (chloroform):
    Ref.: 200, 201

II   R=Bz; 50%; Ref.: 202

III R=Bz; 80%; [α]$_D$ +38.4 (chloroform): Ref.: 198, 202

IV R=Bzl; 80%; [α]$_D$ +8.5 (chloroform): Ref.: 197

V   R=All; 91%; [α]$_D$ -0.5 (chloroform): Ref.: 203

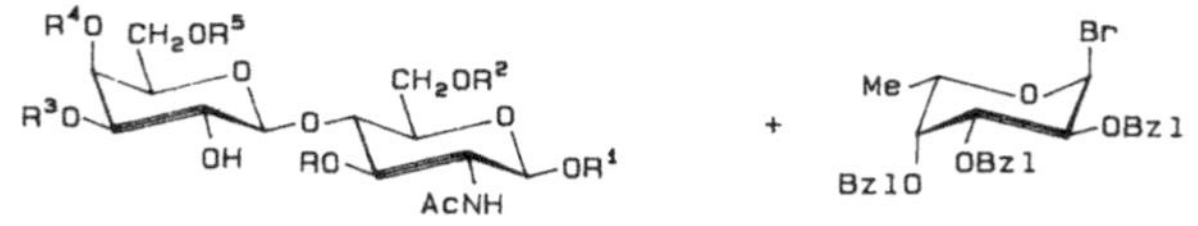

V   Et₄NBr
dichloro-
methane
DMF

VI 1 Et₄NBr
dichloro-
methane
DMF

VII 2 water,
MeOH, AcOH
R⁴=Thp

VIII 1 Et₄NBr
2 H⁺
R³,R⁴=C(Me)₂

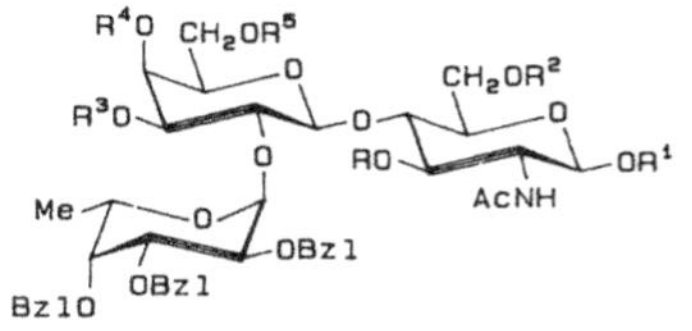

R¹=MCO; R=R²=CH₂OBzl; R³=R⁴=R⁵=Bzl
44%; [α]_D −28.2 (chloroform); C-13; Ref.: 157

R¹=MCO; R=All; R²=CH₂OBzl; R³=R⁴=R⁵=Bzl   VI
94%; C-13; Ref.: 157

R¹=All; R=R²=R³=R⁵=Bzl; R⁴=H   VII
90%; [α]₄₃₆ −58.8 (chloroform); Ref.: 195

R¹=R=R²=Bzl; R³=R⁴=H; R⁵=Ac   VIII
72%; [α]_D −39.8 (chloroform); Ref.: 199

α-L-Fucp-(1→6)-β-D-Galp-(1→4)-D-GlcNAc

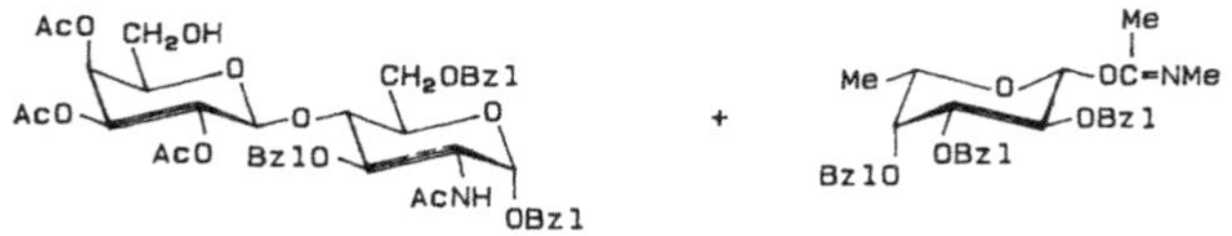

pTSA
benzene

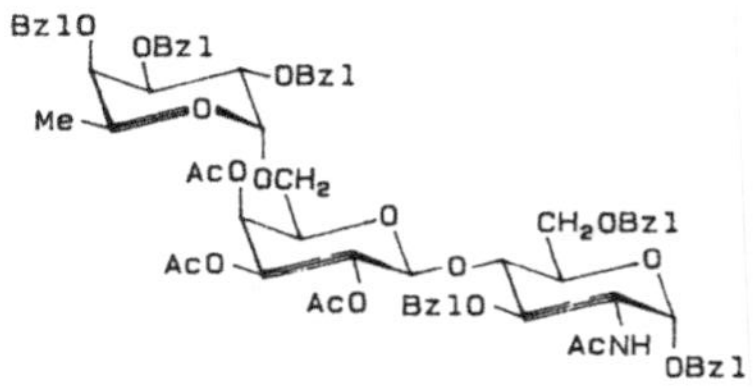

86%; [α]_D +19 (chloroform); Ref.: 204

α-Neu5Ac-(2→6)-β-D-Galp-(1→4)-D-GlcN

α-Neu5Ac-(2→6)-β-D-Galp-(1→4)-D-GlcNAc

$[\alpha]_D$ -0.7 (water): Ref.: 205, 206

Hg(CN)₂
HgBr₂
dichloro-
methane

R¹=Bzl; R=N₃
20%: $[\alpha]_D$ -12.8 (dichloromethane): Ref.: 205, 206

R¹=Ac; R=NPhth
25%: $[\alpha]_D$ +9 (chloroform): Ref.: 207, 208

β-Neu5Ac-(2→6)-β-D-Galp-(1→4)-D-GlcN

β-Neu5Ac-(2→6)-β-D-Galp-(1→4)-D-GlcNAc

$[\alpha]_D$ -3.5 (water): Ref.: 205, 206

Hg(CN)₂
HgBr₂
dichloro-
ethane

R¹=Bzl; R=N₃
20-23%: $[\alpha]_D$ -4.0 (dichloromethane): Ref.: 205, 206

R¹=Ac; R=NPhth
27-37%: $[\alpha]_D$ +21 (chloroform): Ref.: 207, 208

α-D-GlcpN-(1→3)
                    ⟩-D-GlcN
β-D-Ribf-(1→4)

AgClO₄
s-Coll
dibenzo-18-
crown-6
benzene
dioxane

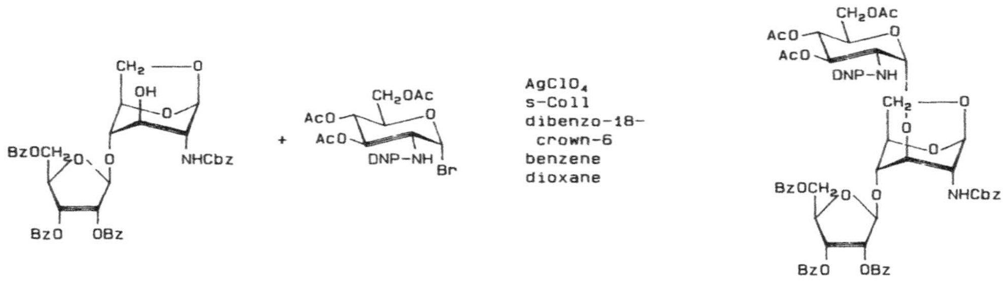

65%; m.p. 99.5-101 °C; [α]_D -3.1; Ref.: 209

β-D-Xylp-(1→3)
                    ⟩-D-GlcNAc
β-D-Galp-(1→4)

white powder; [α]_D +23.8 (water); Ref.: 210

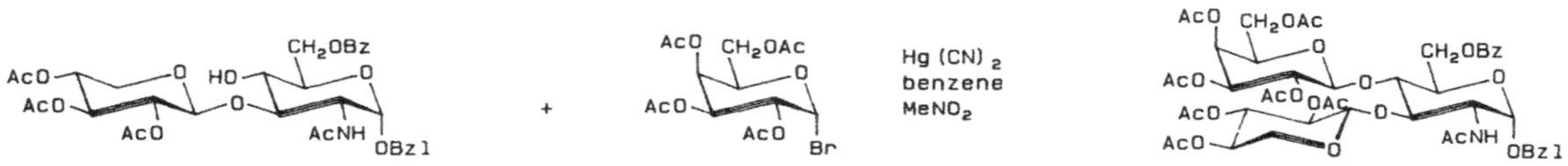

Hg(CN)₂
benzene
MeNO₂

41%; m.p. 226-227.5 °C; [α]_D +14.1 (chloroform); Ref.: 210

α-L-Fucp-(1-3)
                    D-GlcNAc
β-D-GlcpNAc-(1-4)

m.p. 169-172 °C; [α]$_D$ +8.3 → +6.7 (water); C-13; Ref.: 211

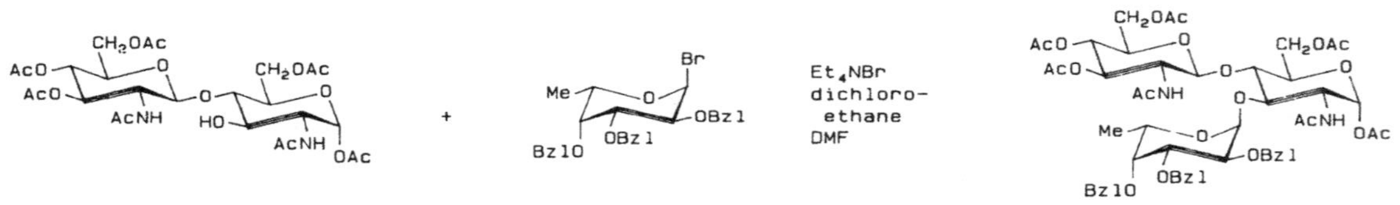

46.6%; m.p. 165-166 °C; [α]$_D$ -39.7 (MeOH); C-13; Ref.: 211

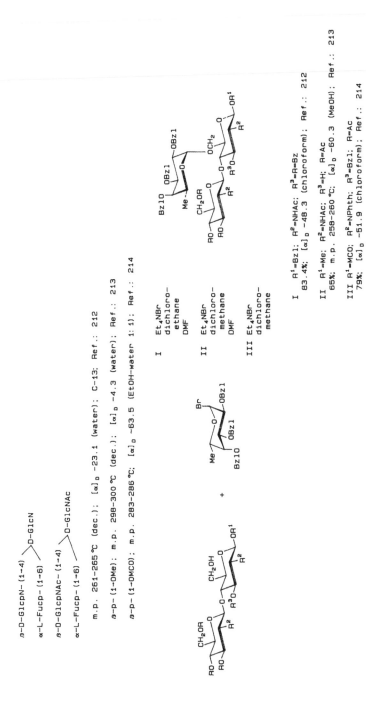

β-D-GlcpN-(1→4)
                    〉D-GlcN
α-L-Fucp-(1→6)

β-D-GlcpNAc-(1→4)
                      〉D-GlcNAc
α-L-Fucp-(1→6)

m.p. 261—265 °C (dec.); [α]$_D$ −23.1 (water); C-13; Ref.: 212

β-D-(1→OMe); m.p. 298—300 °C (dec.); [α]$_D$ −4.3 (water); Ref.: 213

β-D-(1→OMCO); m.p. 283—286 °C; [α]$_D$ −63.5 (EtOH-water 1: 1); Ref.: 214

I   Et$_4$NBr
    dichloro-
    ethane
    DMF

II Et$_4$NBr
    dichloro-
    methane
    DMF

III Et$_4$NBr
    dichloro-
    methane

I   R^1=Bzl; R^2=NHAc; R^3=R=Bz
     83.4%; [α]$_D$ −48.3 (chloroform); Ref.: 212

II   R^1=Me; R^2=NHAc; R^3=H; R=Ac
     65%; m.p. 258—260 °C; [α]$_D$ −60.3 (MeOH); Ref.: 213

III R^1=MCO; R^2=NPhth; R^3=Bzl; R=Ac
     79%; [α]$_D$ −51.9 (chloroform); Ref.: 214

β-D-Galp-(1-3)⟩
β-D-Galp-(1-6)⟩ D-GlcNAc

[α]$_D$ +1.9 (water): Ref.: 215

α-D-(1-OBzl): m.p. 235—236 °C; [α]$_D$ +76.8 (water): Ref.: 215

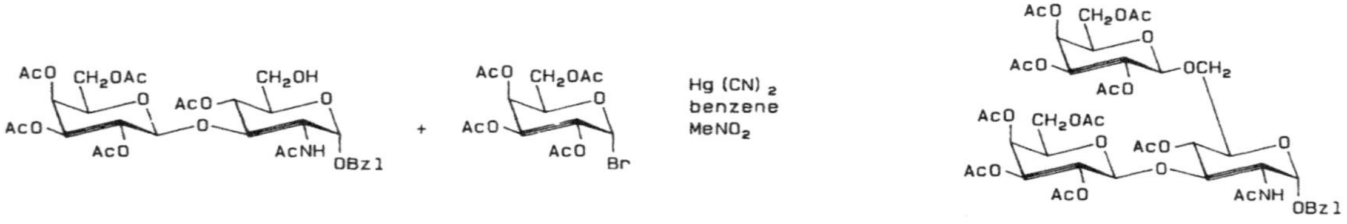

26%; m.p. 230—231 °C; [α]$_D$ +26.3 (chloroform): Ref.: 215

β-D-Galp-(1-3)
α-L-Fucp-(1-4) ⟩ D-GlcN

β-D-Galp-(1-3)
α-L-Fucp-(1-4) ⟩ D-GlcNAc

$[\alpha]_D$ -45.1 (water); C-13; Ref.: 216

$[\alpha]_D$ -44.5 (MeOH-water, 1:19); Ref.: 217

amorphous: $[\alpha]_D$ -43 (water); Ref.: 218

amorphous: $[\alpha]_D$ -44.5 (water); Ref.: 219

$[\alpha]_D$ -43 (water); Ref.: 220

α-(1-OMe); $[\alpha]_D$ -17.4 (water); C-13; Ref.: 219

β-(1-OPh); $[\alpha]_D$ -74.4 (MeOH); Ref.: 221

β-(1-OMCO); $[\alpha]_D$ -73.5 (water); C-13; Ref.: 222

I   Et$_4$NBr, DMF
    (iPr)$_2$EtN
    dichloro-
    methane

II Et$_4$NBr,
    (iPr)$_2$EtN
    dichloro-
    methane

III DPC-ClO$_4$
    AgClO$_4$
    s-Coll
    benzene

IV pTSA
    MeNO$_2$

R^1=H; R^2=OCH$_2$CCl$_3$; R=AC; R^3=Bzl; 83%
m.p. 194-195°C; $[\alpha]_D$ -55.8 (chloroform); Ref.: 216

R^1=H; R^2=OECO; R=Ac; R^3=Bzl; 94%
$[\alpha]_D$ -54 (chloroform); Ref.: 222

R^1=H; R^2=OPh; R=Ac; R^3=Bzl; 73%
m.p. 183-184°C; $[\alpha]_D$ -66.2 (chloroform); Ref.: 221

R^1=H; R^2=OBzl; R=Bzl; R^3=Bzl; 82.7%; m.p. 112-114°C
$[\alpha]_D$ -74.9; (chloroform); C-13; Ref.: 219

R^1=OMe; R^2=H; R=Bzl; R^3=Bzl; 86%
$[\alpha]_D$ +3.2 (chloroform); Ref.: 219

II R^1=OBzl; R^2=H; R=Ac; R^3=Bzl; 60%
    m.p. 202—203 °C; $[\alpha]_D$ +14 (chloroform): Ref.:: 220

IV R^1=OBzl; R^2=H; R=Bzl; R^3=Bzl; 85%
    $[\alpha]_D$ +15.5 (chloroform): Ref.:: 217

III R^1=OBzl; R^2=H; R=Ac; R^3=NBz; 34%; m.p. 227 °C;
    $[\alpha]_D$ -100 (chloroform): Ref.:: 218

    R^1=OBzl; R^2=H; R=Ac; R^3=NBz; 43%; m.p. 227 °C;
    $[\alpha]_D$ -100 (chloroform): Ref.:: 223

    R^1=OBzl; R^2=H; R=Ac; R^3=NBz; 60%; m.p. 227 °C;
    $[\alpha]_D$ -100 (chloroform): Ref.:: 220

β-D-Galp-(1→3)
                  ⟩D-GlcNAc
α-L-Fucp-(1→6)

white, amorphous powder: $[\alpha]_D$ -43 (water): C-13: Ref.:: 216

45%; m.p. 144—145 °C: $[\alpha]_D$ -26.5 (chloroform): Ref.:: 216

β-D-Galp-(1→4) ⟩D-GlcNAc
β-D-Galp-(1→6)

α-D-(1-OBz1); m.p. 160-161 °C; [α]$_D$ +86 (water); Ref.: 155

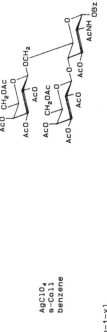

DCP = 2,3-diphenyl-2-cyclopropen-1-yl

[α]$_D$ +30 (chloroform); Ref.: 155

α-L-Fucp-(1→3)
β-D-Galp-(1→4) ⟩D-GlcN

α-L-Fucp-(1→3)
β-D-Galp-(1→4) ⟩D-GlcNAc

powder; m.p. 140—142 °C; [α]_D −33 (water); Ref.: 224

β-p-(1-OMCO); [α]_D −64 (water); C-13; Ref.: 213

I   Et₄NBr
    DMF
    dichloro-
    ethane

II Et₄NBr
    DMF
    dichloro-
    methane

III AgOTf
    s-Coll
    dichloro-
    methane
    toluene

I   R¹=OBzl; R²=H; R³=Bzl; R⁴=R=Ac; R⁵=NHAc
     55%; m.p. 154—155 °C; [α]_D +6.5 (chloroform); Ref.: 224

     R¹OBzl; R²=H; R³=R⁴=R=Ac; R⁵=NHAc
     52%; m.p. 132—133 °C; [α]_D +15 (chloroform); Ref.: 224

II   R¹=H; R²=OMCO; R³=CH₂OCH₂CCl₃; R⁴=Bz;
     R=Bzl; R⁵=NHAc
     80%; [α]_D −33 (chloroform); Ref.: 213

III R¹=H; R²=SEt; R³=Bzl; R⁴=R=Ac; R⁵=NPhth
     53%; m.p. 169 °C; [α]₅₇₈ +6 (chloroform); C-13; Ref.: 225

α-L-Fucp-(1→4)
                    D-GlcN
α-L-Fucp-(1→6)

α-L-Fucp-(1→4)
                    D-GlcNAc
α-L-Fucp-(1→6)

α-p-(1-OMe): colourless, amorphous solid; $[\alpha]_D$ -3.8 (water); Ref.: 213

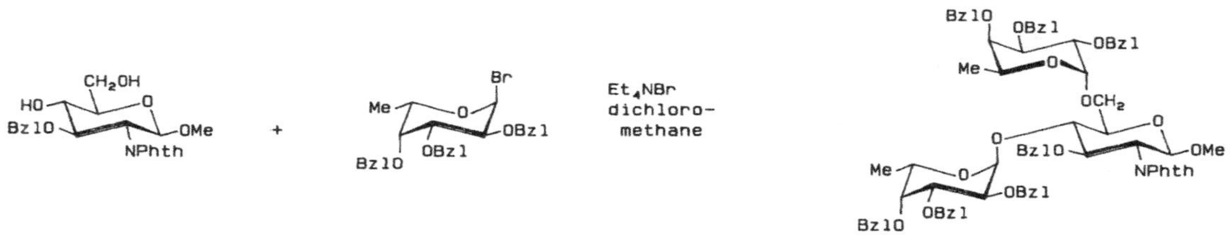

$[\alpha]_D$ -38.2 (chloroform); Ref.: 213

β-D-Galp-(1→4)-β-D-GlcpN-(1→2)-D-Man

β-D-Galp-(1→4)-β-D-GlcpNAc-(1→2)-D-Man

[α]_D -13 (water): Ref.: 226

amorphous powder; [α]_D -22 (water); C-13; Ref.: 227

α-p-(1-OPNP); amorphous powder; [α]_D +38 (water); C-13; Ref.: 228

α-p-(1-OPr); [α]_D -4.5 (MeOH); C-13; Ref.: 229

I   pTSA
    toluene
    MeNO_2

II AgOTf
    s-Coll
    dichloro-
    methane

II

III AgOTf
    dichloro-
    ethane

I   8.6%; [α]_D -26 (chloroform); Ref.: 226

II R¹=R=Bzl; R², R²=PhCH
    66%; m.p. 167–168 °C; [α]_D +19 (chloroform);
    C-13; Ref.: 227

    R¹=PNP; R=Bz; R², R²=PhCH
    60%; [α]_D +53 (chloroform); C-13; Ref.: 228

    R¹=Me; R=R²=Bzl
    83%; m.p. 133 °C; [α]_D -11.7 (chloroform); Ref.: 230

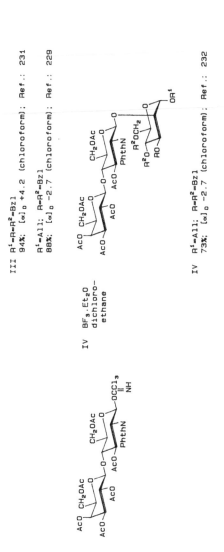

III R¹=R=R²=Bzl
94%: [α]_D +4.2 (chloroform): Ref.: 231

R¹=All; R=R²=Bzl
88%: [α]_D -2.7 (chloroform): Ref.: 229

IV R¹=All; R=R²=Bzl
73%: [α]_D -2.7 (chloroform): Ref.: 232

IV BF₃·Et₂O
dichloro-
ethane

α-L-Fucp-(1→3)-β-D-GlcpN-(1→2)-D-Man

α-L-Fucp-(1→3)-β-D-GlcpNAc-(1→2)-D-Man

[α]_578 -100 (water): C-13; Ref.: 233

MeOTf
Et₂O

72%: [α]_578 -6 (chloroform): C-13: Ref.: 233

β-D-Galp-(1→4)-β-D-GlcpNAc-(1→3)-D-Man

amorphous; [α]$_D$ -14 (water); Ref.: 234

pTSA
dichloro-
ethane

9%; m.p. 120—121°C; [α]$_D$ +8 (chloroform); Ref.: 234

β-D-Galp-(1→4)-β-D-GlcpN-(1→4)-D-Man

β-D-Galp-(1→4)-β-D-GlcpNAc-(1→4)-D-Man

α-p-(1-OPr); [α]$_D$ +24.5 (water); C-13; Ref.: 235

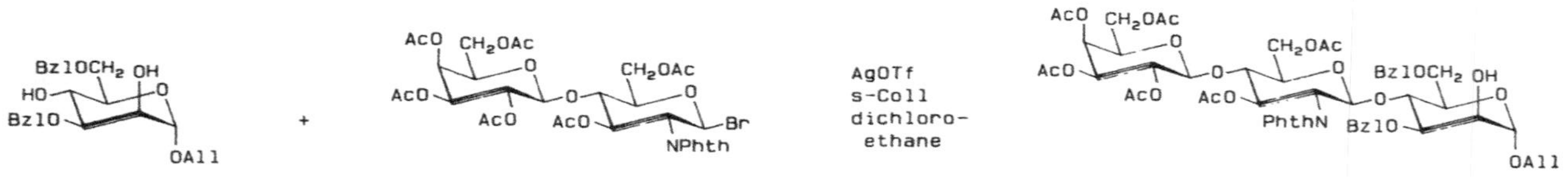

AgOTf
s-Coll
dichloro-
ethane

41.7%; [α]$_D$ +29 (chloroform); C-13; Ref.: 235

β-D-Galp-(1→3)-β-D-GlcpNAc-(1→6)-D-Man

α-p-(1→OPNP); [α]$_D$ -37.1 (water); C-13; Ref.: 236

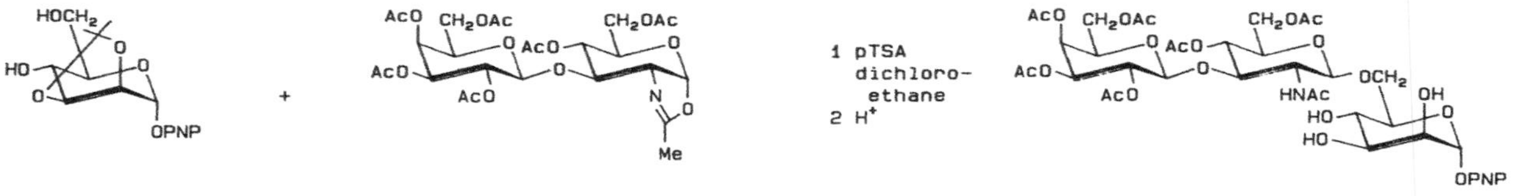

22%; [α]$_D$ +41.2 (dichloromethane); C-13; Ref.: 236

β-D-Galp-(1→4)-β-D-GlcpN-(1→6)-D-Man

β-D-Galp-(1→4)-β-D-GlcpNAc-(1→6)-D-Man

amorphous solid: [α]$_D$ -12 (water): Ref.: 237

amorphous powder: [α]$_D$ -10 (water): C-13: Ref.: 238

α-p-(1-OPNP): amorphous; [α]$_D$ +41.6 (water): C-13: Ref.: 236

I   AgOTf
    MeNO$_2$

II pTSA
    dichloro-
    ethane

III pTSA
    dichloro-
    methane

I   R^1=R^2=Bzl: R^3=NPhth;   [α]$_D$ +29 (chloroform):
    63%: m.p. 146-147°C;
    C-13: Ref.: 238

II R^1=R^2=Bzl: R^3=NHAc
    52%: [α]$_D$ +9 (chloroform): Ref.: 237

III R^1=PNP: R, R=C(Me)$_2$: R^3=NHAc
    53%: [α]$_D$ +18.7 (chloroform): Ref.: 236

α-D-Manp-(1→2)-α-D-Manp-(1→2)-D-Man

m.p. 183-185 °C; [α]$_D$ +55.3 (water); C-13; Ref.: 239

amorphous; [α]$_D$ +48 (water); C-13; Ref.: 240

α-p-(1-OPr); [α]$_D$ +31 (water); C-13; Ref.: 241

α-O—CH$_2$CH—OC$_{14}$H$_{29}$; [α]$_D$ +60.6 (THF); C-13; Ref.: 242
          |
        OC$_{14}$H$_{29}$

α-(1-DMCO) 6'-disodiumphosphate, white powder; [α]$_D$ +41.5 (water); C-13; Ref.: 243

I   AgOTf
      dichloro-
      ethane

II   Hg(CN)$_2$
      dichloro-
      ethane

III 1 AgOTf
       dichloro-
       methane
     2 NaOMe

IV   1 AgOTf
       dichloro-
       ethane
     2 NaOMe

II
V   Hg(CN)$_2$
     benzene
     MeNO$_2$

I   R^1=Bzl; R=R^2=Ac; No data; Ref.: 239

II   [1+2] R^1=All; R=R^2=Bzl; X=Cl;
      [α]$_D$ +17 (chloroform); C-13; Ref.: 241

II   [2+1] R^1=All; R=R^2=Bzl; X=Cl; 49%; Ref.: 241

II   [2+1] R^1=All; R=R^2=Bzl; X=Br; 50%; Ref.: 241

V   [2+1] R^1=All; R=R^2=Bzl; X=Cl; 58%; Ref.: 241

IV   R^1=R=Bzl; R^2=H; 71%;
      [α]$_D$ +34.2 (chloroform); Ref.: 239

III   R^1=—OC$_{14}$H$_{29}$; R=Bzl; R^2=H; 42%;
             |
           OC$_{14}$H$_{29}$

      [α]$_D$ +32.6 (chloroform); Ref.: 242

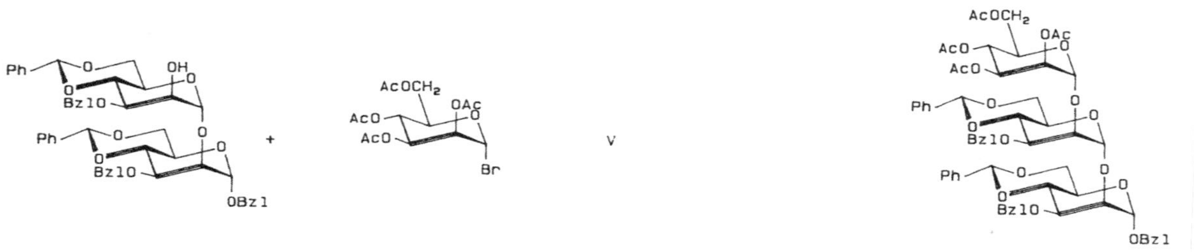

V   67%; [α]_D +34 (chloroform); C-13; Ref.: 240

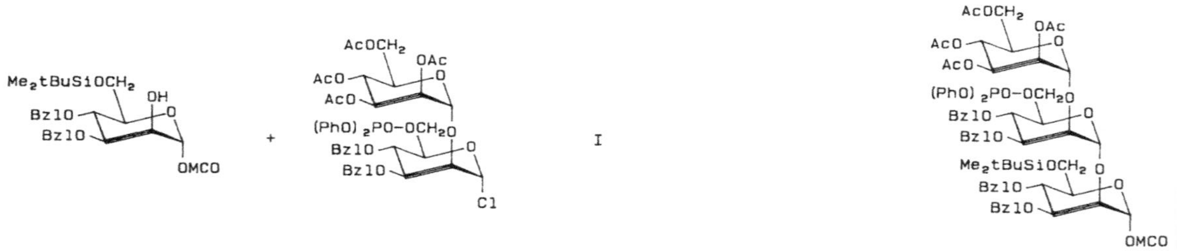

I   45%; syrup; [α]_D +28 (chloroform); C-13; Ref.: 243

α-D-Manp-(1→2)-α-D-Manp-(1→2)-D-Man

α-D-(1→OPr): $[\alpha]_D$ +13 (water); C-13; Ref.: 241

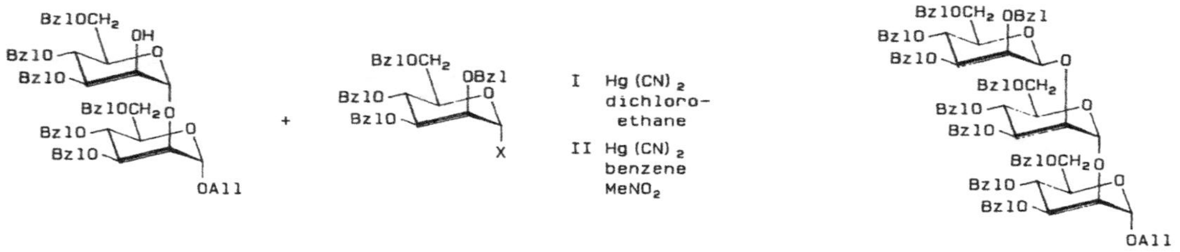

I   Hg(CN)$_2$ dichloro-ethane

II Hg(CN)$_2$ benzene MeNO$_2$

$[\alpha]_D$ -26 (chloroform); C-13; Ref.: 241

I   13%; X=Cl

~ 50%; X=Br

II 40%; X=Cl

β-D-GlcpN-(1→2)-α-D-Manp-(1→3)-D-Man

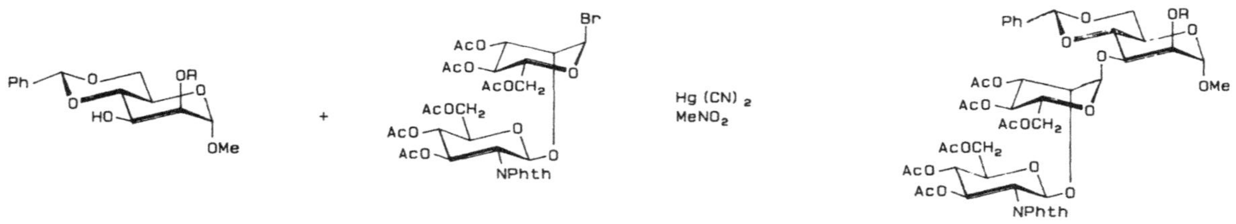

Hg(CN)₂
MeNO₂

R=Ac: 49.4%; [α]_D +12.9 (chloroform); C-13; Ref.: 244
R=All: 52.2%; [α]_D -4.6 (chloroform); C-13; Ref.: 244

α-D-Manp-(1→2)-α-D-Manp-(1→3)-D-Man

α-p-(1-OMCO) 6'-disodiumphosphate; white powder; [α]_D +60.6 (water); C-13; Ref.: 243

AgOTf
s-Coll
dichloro-
ethane

51%; syrup: [α]_D +34.4 (chloroform); C-13; Ref.: 243

α-D-Manp-(1→3)-α-D-Manp-(1→3)-D-Man

amorphous powder: [α]$_D$ +19.3 (water); C-13; Ref.: 245

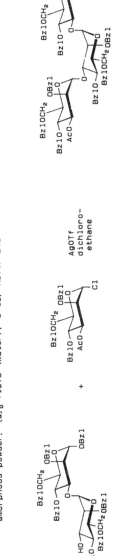

AgOTf
dichloro-
ethane

91%; [α]$_D$ +14 (chloroform); Ref.: 245

β-D-GlcpN-(1→2)-α-D-Manp-(1→6)-D-Man

β-D-GlcpNAc-(1→2)-α-D-Manp-(1→6)-D-Man

β-p-(1→OMCO); [α]$_D$ -19.3 (water); C-13; Ref.: 246

AgOTf
s-Coll
dichloro-
methane

76%; [α]$_D$ -23.0 (chloroform); C-13; Ref.: 246

β-D-GlcpN-(1→6)-α-D-Manp-(1→6)-D-Man

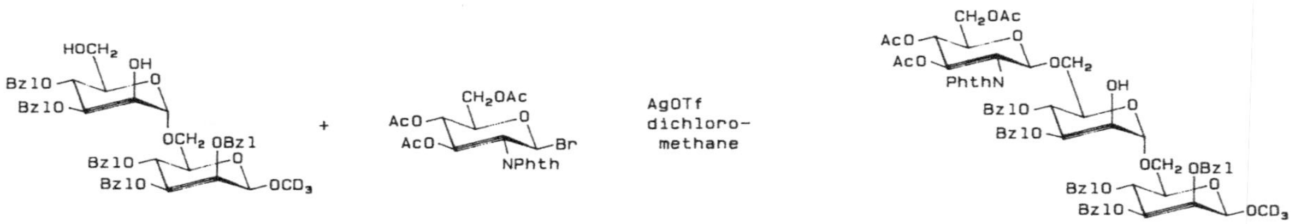

AgOTf
dichloro-
methane

71%; [α]_D +3.0 (chloroform); Ref.: 247

α-D-Manp-(1→2)-α-D-Manp-(1→6)-D-Man

α-p-(1-OMCO) 6'-disodiumphosphate; [α]_D +53.2 (water); C-13; Ref.: 243

1 AgOTf
  s-Coll
  4 Å
  dichloro-
  ethane

2 Pd/C; H₂

3 Ac₂O; Py

19%; syrup; [α]_D +44.2 (chloroform); C-13; Ref.: 243

α-D-Manp-(1→2)-β-D-Manp-(1→2)-D-Man

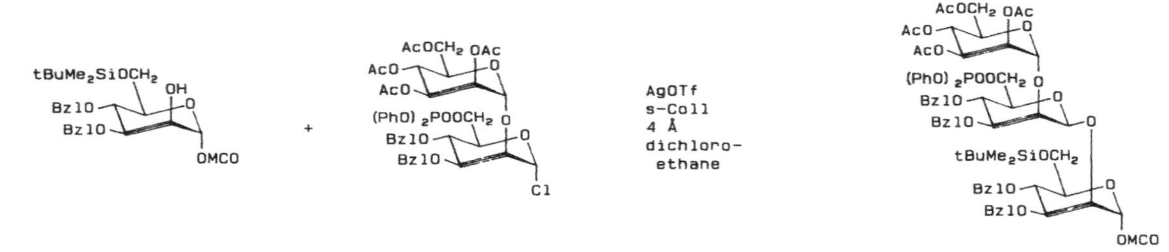

37%; [α]_D +7.0 (chloroform): C-13; Ref.: 243

α-D-Manp-(1→2)-β-D-Manp-(1→3)-D-Man

22%; syrup: [α]_D 0 (chloroform): C-13; Ref.: 243

α-D-Manp-(1→2)-β-D-Manp-(1→6)-D-Man

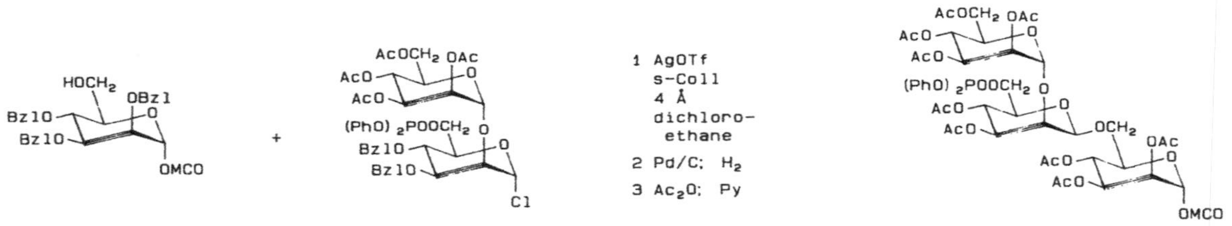

1 AgOTf
  s-Coll
  4 Å
  dichloro-
  ethane
2 Pd/C; H₂
3 Ac₂O; Py

57%; [α]D +16.4 (chloroform); C-13; Ref.: 243

β-D-Xylp-(1→2)
                 ⟩D-Man
β-D-Xylp-(1→3)

α-D-(1-OBzl): m.p. 123-125 °C; [α]D -8.3 (water); Ref.: 248
α-D-(1-OMe): hygroscopic: m.p. 122-124 °C; [α]D -21.7 (water); Ref.: 248

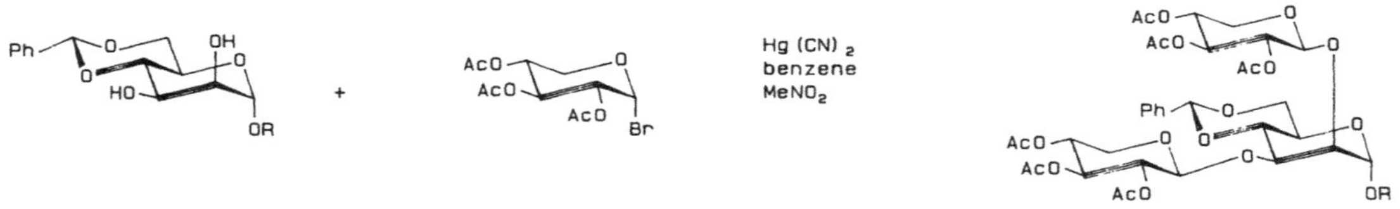

Hg(CN)₂
benzene
MeNO₂

R=Bzl; 37%; m.p. 92-94 °C; Ref.: 248
R=Me; 6%; m.p. 108 °C; Ref.: 248

α-D-Lyxp-(1→3)
                    >D-Man
α-D-Manp-(1→6)

α-D-(1→OMe): [α]_D +91.5 (water): Ref.: 249

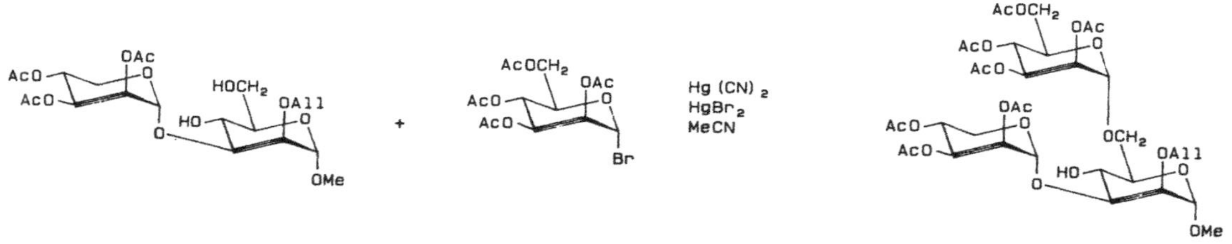

58%; Ref.: 249

α-D-Manp-(1→3)
                    >D-Man
α-D-Lyxp-(1→6)

α-D-(1→OMe): [α]_D +89.6 (water): Ref.: 249

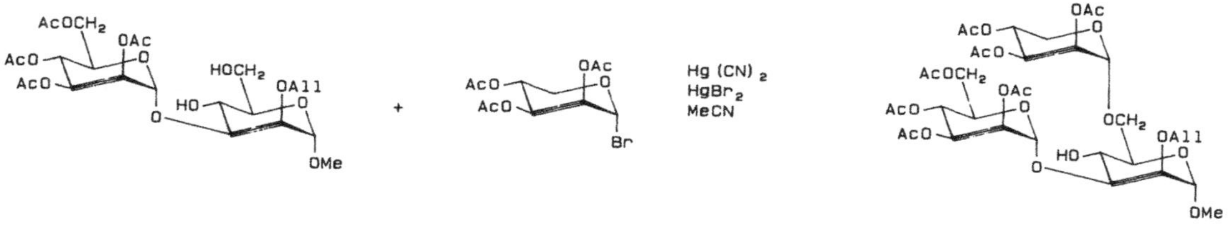

59%; Ref.: 249

β-D-GlcpN-(1→2)
β-D-GlcpN-(1→4) ⟩D-Man

β-D-GlcpNAc-(1→2)
β-D-GlcpNAc-(1→4) ⟩D-Man

amorphous powder: $[\alpha]_D$ −16 (water); C-13; Ref.: 250

α-D-(1→OMe); $[\alpha]_D$ +9.1 (MeOH); C-13; Ref.: 251

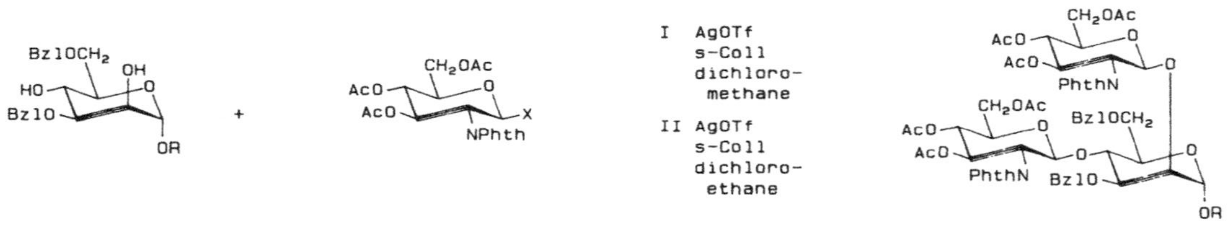

I AgOTf
s-Coll
dichloro-
methane

II AgOTf
s-Coll
dichloro-
ethane

I R=Bzl; X=Br; 54%; $[\alpha]_D$ +26 (chloroform); C-13; Ref.: 250

II R=Me; X=Cl; 60%; $[\alpha]_D$ +12.5 (chloroform); Ref.: 251

β–D–GlcpN–(1→2) ＼
＞D—Man
β–D–GlcpN–(1→6) ／

β–D–GlcpNAc–(1→2) ＼
＞D—Man
β–D–GlcpNAc–(1→6) ／

amorphous powder: [α]$_D$ –33 (water); C–13; Ref.: 250

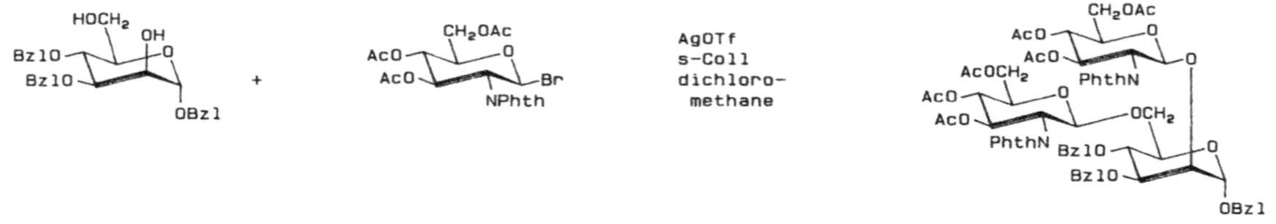

AgOTf
s-Coll
dichloro-
methane

32%; [α]$_D$ +9 (chloroform); C–13; Ref.: 250

β-D-GlcpN-(1→3)
β-D-GlcpN-(1→6) ⟩D-Man
β-D-GlcpNAc-(1→3)
β-D-GlcpNAc-(1→6) ⟩D-Man

α-D-(1→OMe): [α]$_D$ −6.9 (water): C−13: Ref.: 251

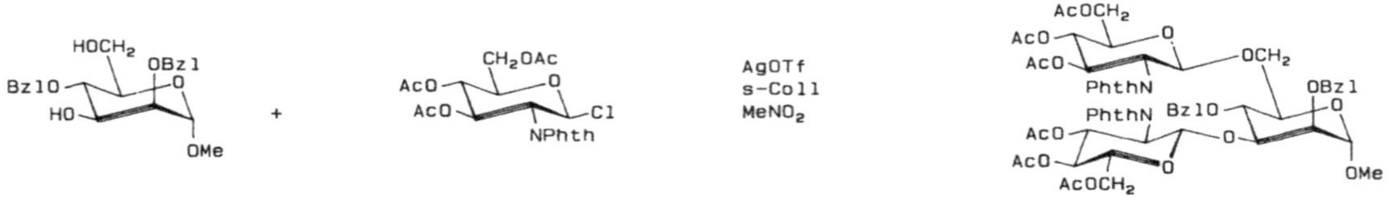

AgOTf
s-Coll
MeNO$_2$

41%: [α]$_D$ +5.8 (chloroform): Ref.: 251

α-D-Manp-(1→2)
α-D-Manp-(1→4) ⟩D-Man

α-D-(1→OMe): [α]$_D$ +38.5 (water): C−13: Ref.: 252

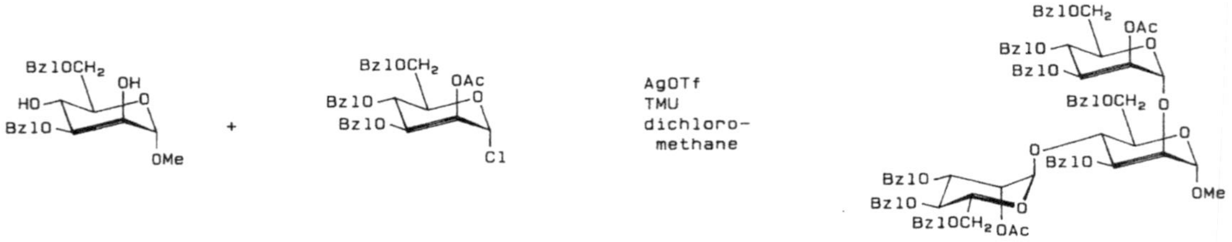

AgOTf
TMU
dichloro-
methane

69.7%: [α]$_D$ +23.5 (chloroform): C−13: Ref.: 252

α-D-Manp-(1→2)
                    〉D-Man
α-D-Manp-(1→6)

α-D-(1→OMe): m.p. 207—208 °C; $[\alpha]_D$ +76.6 (water); C-13; Ref.: 253

α-D-(1→OMe): amorphous; $[\alpha]_D$ +66.7 (water); C-13; Ref.: 252

I   dichloro-
     methane

II   AgOTf
     TMU
     dichloro-
     methane

III Hg(CN)$_2$
     MeNO$_2$

I R^1=(CH$_2$)$_2$-Ph-NHTs; R^2=R=Bzl; R^3=Bz
     74%; $[\alpha]_D$ +14.4; C-13; Ref.: 254

II R^1=Me; R^2=R=Bzl; R^3=Ac
     75.2%; $[\alpha]_D$ +30.7 (chloroform); C-13; Ref.: 252

III R^1=Me; R^2=H; R=R^3=Ac
     60%; m.p. 67—68 °C; $[\alpha]_D$ +47.7 (chloroform);
     C-13; Ref.: 253

α-D-Manp-(1→3) ⟩ D-Man
α-D-Manp-(1→6)

[α]$_D$ +59 (water): Ref.: 255

α-D-(1→OMe): [α]$_D$ +96.7 (MeOH): Ref.: 256, 257, 258
α-D-(1→OMe): [α]$_D$ +93.6 (water): Ref.: 259
α-D-(1→OMe): [α]$_D$ +83.9 (water): Ref.: 260
α-D-(1→OMe): [α]$_D$ +111 (MeOH): Ref.: 261
β-D-(1→OECO): [α]$_D$ +24.3 (water): C-13; Ref.: 229
α-D-(1→O-$C_{14}H_{29}$): [α]$_D$ +62.0 (THF): C-13; Ref.: 242

I   1 dichloro-
        ethane
    2 BzlBr
    3 HgBr$_2$

II AgOTf; TMU
    dichloro-
    methane

III AgOTf; TMU
    dichloro-
    ethane

IV 1 Hg(CN)$_2$
       HgBr$_2$
    2 NaOMe

I   R=Bu$_3$Sn; R^1=Me; R^2=Ac
     [α]$_D$ +41.7 (chloroform): Ref.: 256, 257

II   R=Bzl; R^1=Me; R^2=Ac
     79%: [α]$_D$ +41.7 (chloroform): Ref.: 257
     4.5%: [α]$_D$ +46.1 (chloroform): Ref.: 259

III R=Bzl; R^2=Ac; R^1=—OC$_{14}$H$_{29}$: Ref.: 242
                              OC$_{14}$H$_{29}$

IV   R=R^1=Bzl; R^2=H
     49%: [α]$_D$ +58 (chloroform): Ref.: 255

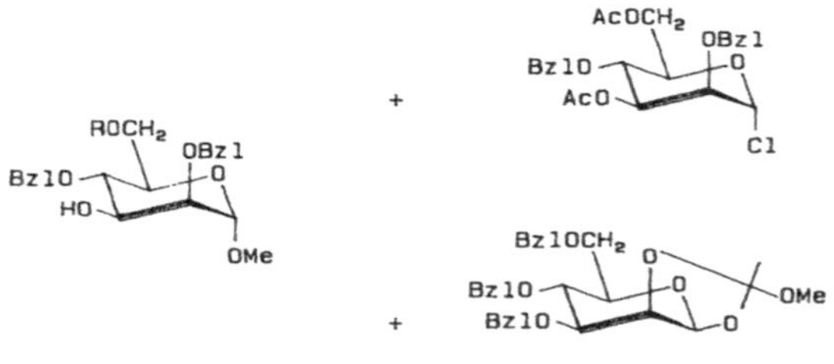

II

V HgBr₂

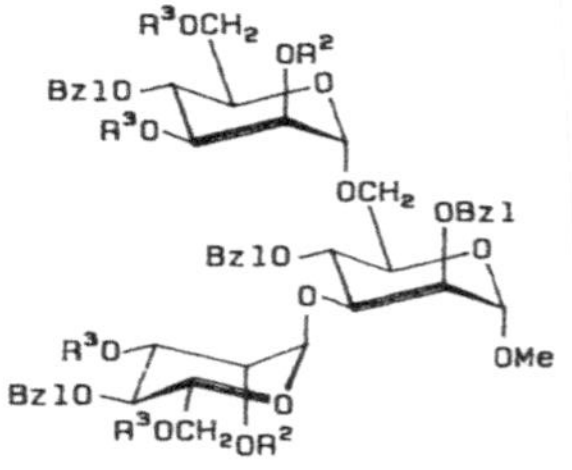

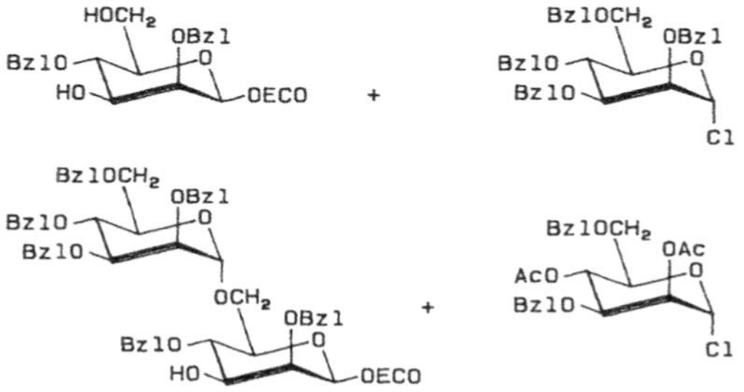

VI Bu₄NCl dichloro-ethane

VII AgOTf dichloro-ethane

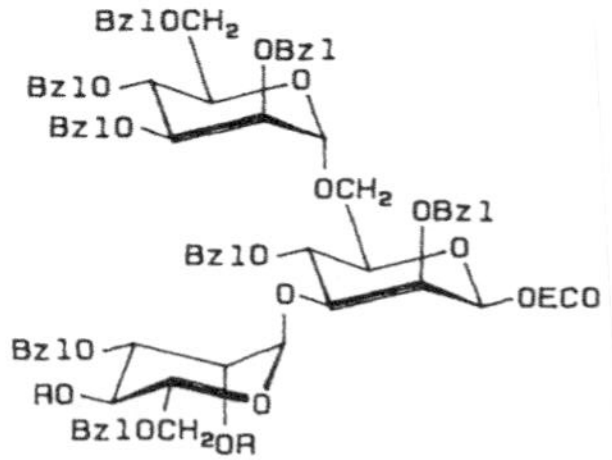

II R=tBu(Me)₂Si; R²=Bzl; R³=Ac; 40.9%; Ref.: 259

R=H; R²=Bzl; R³=Ac; 20%; [α]$_D$ +36.6; Ref.: 258

V R=H; R²=Ac; R³=Bzl; 24%; Ref.: 257

VI R=Bzl; 10.2%; [α]$_D$ +9.8 (chloroform); C-13; Ref.: 229

VII R=Ac; 78%; [α]$_D$ -3.1 (chloroform); C-13; Ref.: 267, 268

VIII Hg(CN)₂ HgBr₂ MeCN

IX   HgBr₂ MeCN

X   AgOTf dichloro-methane

II   AgOTf TMU dichloro-methane

III AgOTf TMU dichloro-ethane

II   AgOTf s-coll dichloro-methane

XI   AgOTf s-coll dichloro-methane

VIII $R=R^3=R^4=R^5=R^6=R^7=Ac$; $R^1=All$; $R^2=H$
67%; $[\alpha]_D$ +53.4 (chloroform): $R^2=H$; Ref.: 261

IX $R=R^3=R^4=R^5=R^6=R^7=Ac$; $R^1=All$; $R^2=H$; Ref.: 262

X $R=R^1=R^2=R^4=R^5=Bzl$; $R^3=R^6=R^7=Ac$
82.2%; $[\alpha]_D$ +43.2 (chloroform); C-13; Ref.: 260

II $R=R^1=R^2=R^3=R^5=Bzl$; $R^4=R^6=R^7=Ac$; 91%; Ref.: 258

III $R=R^1=R^2=R^4=R^6=Bzl$; $R^3=R^5=R^7=Ac$
62%; $[\alpha]_D$ +35.2 (chloroform): Ref.: 263, 264

II $R=Me$; $R^1=R^2=Bzl$; $R^3=Ac$
83.6%; $[\alpha]_D$ +41.3 (chloroform); C-13; Ref.: 259

XI $R=R^3=Bzl$; $R^1=R^2=Ac$
59%; $[\alpha]_{578}$ +33 (chloroform); C-13; Ref.: 265

$R=R^1=R^3=Bzl$; $R^2=Ac$
47%; $[\alpha]_D$ +37 (chloroform); C-13; Ref.: 266

α–D–Galp–(1→2)

3,6-Dideoxy-α-D-ribo-Hexp-(1→3) ⟩D-Man

α–p–(1→OMCO): [α]_D +143.8 (water); C-13; Ref.: 269

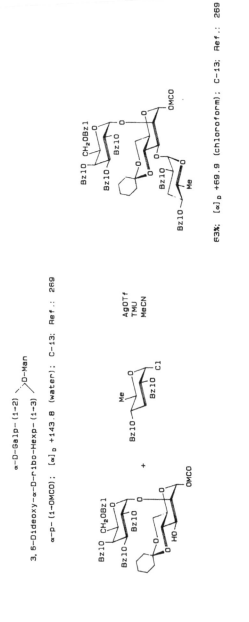

AgOTf
TMU
MeCN

63%; [α]_D +69.9 (chloroform); C-13; Ref.: 269

α–D–Galp–(1→2)

3,6-Dideoxy-α-D-arabino-Hexp-(1→3) ⟩D-Man

α–p–(1→OMCO): [α]_D +120.3 (water); C-13; Ref.: 269

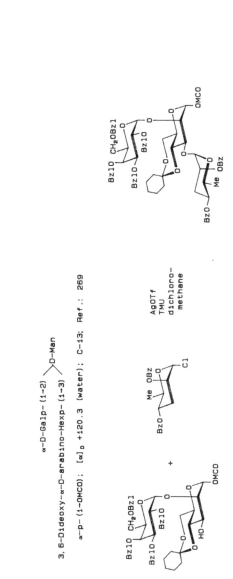

AgOTf
TMU
dichloro-
methane

67%; [α]_D +49.9 (chloroform); C-13; Ref.: 269

α-D-Galp-(1→2)
                    D-Man
3,6-Dideoxy-α-D-xylo-Hexp-(1→3)

α-p-(1→OMCO): [α]ᴅ +106.8 (water); C-13; Ref.: 270

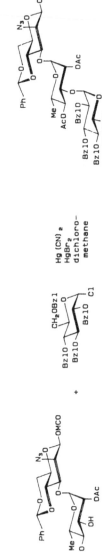

67%; [α]ᴅ +95.2 (dichloromethane); C-13; Ref.: 270

α-D-Glcp-(1→3)-α-L-Rhap-(1→3)-D-ManN
α-D-Glcp-(1→3)-α-L-Rhap-(1→3)-D-ManNAc

β-p-(1→OMCO): [α]ᴅ -1.2 (MeOH); Ref.: 271

54%; Ref.: 271

α-L-Rhap-(1→3)-α-D-GlcpNAc-(1→2)-L-Rha

α-p-(1-OMCO): [α]_D -46.4 (water); C-13; Ref.: 272

α-p-(1-O-8-Hydrazinocarbonyloctyl): [α]_D -47.8 (water); C-13; Ref.: 272

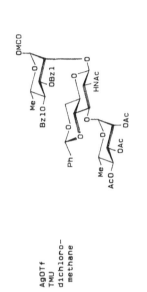

AgOTf
TMU
dichloro-
methane

73%; [α]_D -34.3 (chloroform); C-13; Ref.: 272

α-D-Galp-(1→2)-α-D-Manp-(1→4)-L-Rha

α-p-(1-O-p-Trifluoroacetamidophenyl); [α]_D +36 (chloroform); C-13; Ref.: 273

I   AgOTf
    s-Coll
    toluene

II AgOTf
    MeNO_2
    toluene

I   R=(S)-PhCH; R¹=p-Nitrophenyl; R²=R³=R⁴=Bzl
    76%; [α]_D +16 (chloroform); C-13; Ref.: 273

II R=cHex; R¹=MCO; R²=All; R³=Bz; R⁴=Ac
    88%; [α]_D +63 (chloroform); Ref.: 274

α-D-Galp-(1→2)-β-D-Manp-(1→4)-L-Rha

α-p-(1-OMCO): [α]$_D$ +0.2 (water); C-13; Ref.: 275

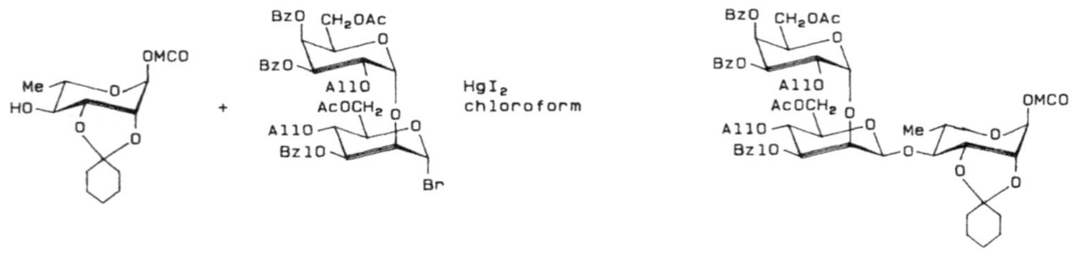

26%; [α]$_D$ +70 (chloroform); Ref.: 275

β-D-Galp-(1→6)-β-D-Manp-(1→4)-L-Rha

α-p-(1-O-p-Trifluoroacetamidophenyl); [α]$_D$ -68 (water); C-13; Ref.: 276

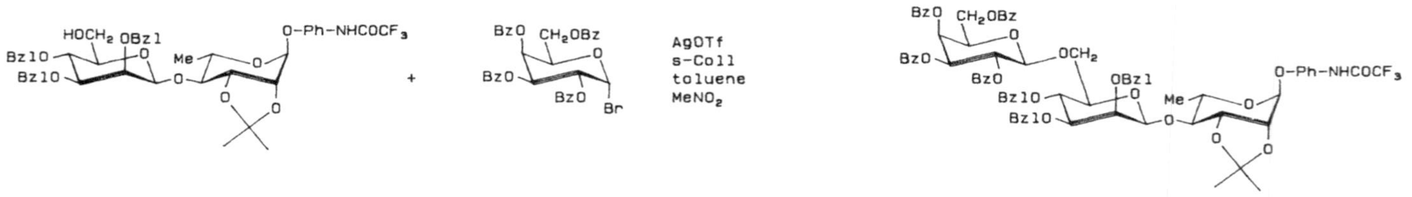

84%; [α]$_D$ -16 (chloroform); C-13; Ref.: 276

α-D-Tyvp-(1→3)-β-D-Manp-(1→4)-L-Rha

C-13; Ref.: 277

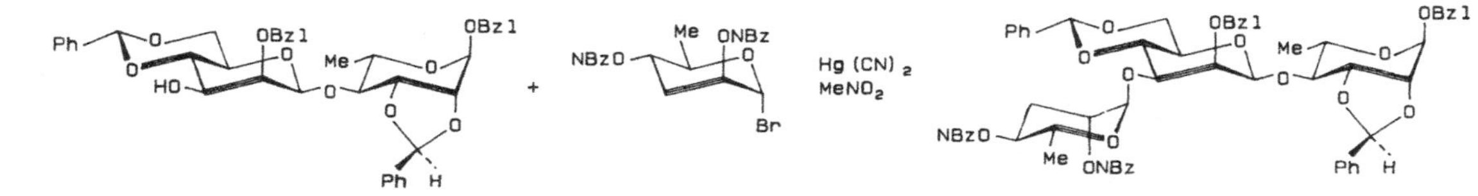

35%; Ref.: 277

α-D-Glcp-(1→3)-α-L-Rhap-(1→2)-L-Rha

α-p-(1→OMe): amorphous; [α]_D +34 (MeOH); C-13; Ref.: 278

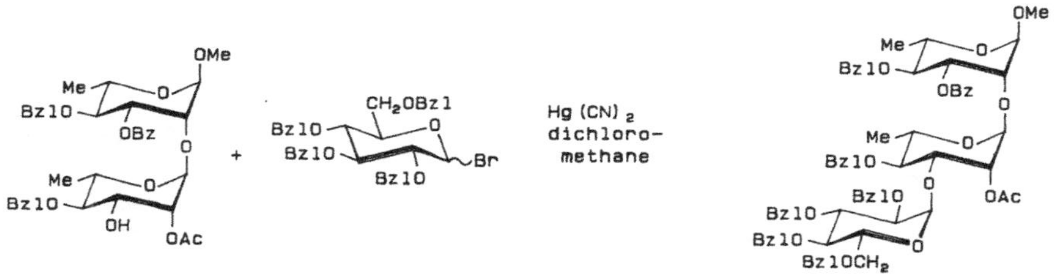

[α]_D +45 (chloroform); C-13; Ref.: 278

α-D-Glcp-(1→3)-α-L-Rhap-(1→2)-L-Rha

1 Hg(CN)₂ dichloro-
  methane

2 NaOMe

17%: [α]$_D$ -13 (chloroform); C-13; Ref.: 278

3,6-di-O-Me-β-D-Glcp-(1→4)-2,3-di-O-Me-α-L-Rhap-(1→2)-3-O-Me-L-Rha

α-p-(1-OPr); oil; [α]$_D$ -47.4 (chloroform); Ref.: 279

Hg(CN)₂
MeCN

Hg(CN)₂
MeCN

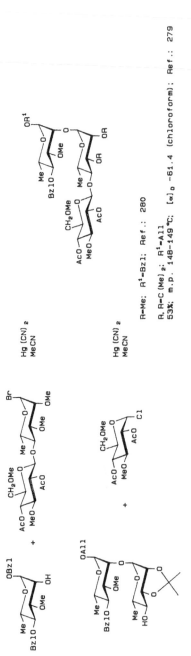

R=Me; R¹=Bzl; Ref.: 280

R,R=C(Me)₂; R¹=All
53%; m.p. 148-149 °C; [α]$_D$ -61.4 (chloroform); Ref.: 279

β-D-GlcpN-(1→2)-α-L-Rhap-(1→2)-L-Rha
β-D-GlcpNAc-(1→2)-α-L-Rhap-(1→2)-L-Rha

α-p-(1→OMCO): [α]589 -36.4 (MeOH): C-13: Ref.: 281

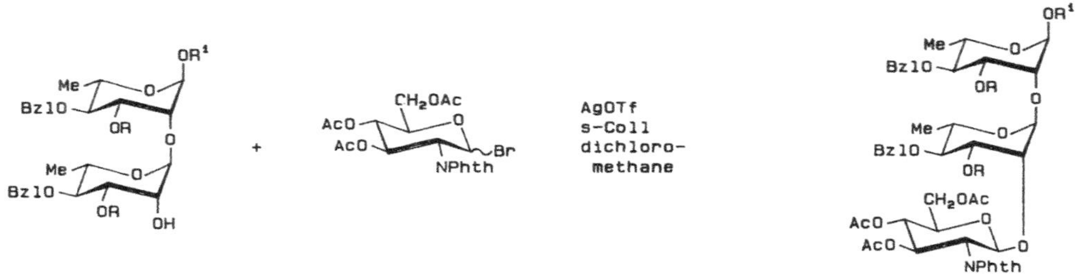

AgOTf
s-Coll
dichloro-
methane

R=Bzl; R1=MCO
70%; m.p. 117-118°C; [α]589 +6.1 (chloroform): C-13: Ref.: 281

R=All; R1=Me
75%; [α]D +3.6 (dichloromethane): C-13: Ref.: 282

α-D-Rhap-(1→2)-α-L-Rhap-(1→2)-L-Rha

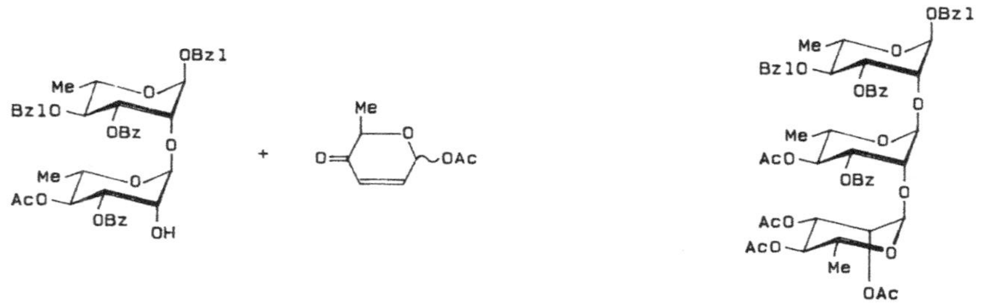

[α]D +42.2 (chloroform): C-13: Ref.: 283

α-L-Rhap-(1→2)-α-L-Rhap-(1→2)-L-Rha

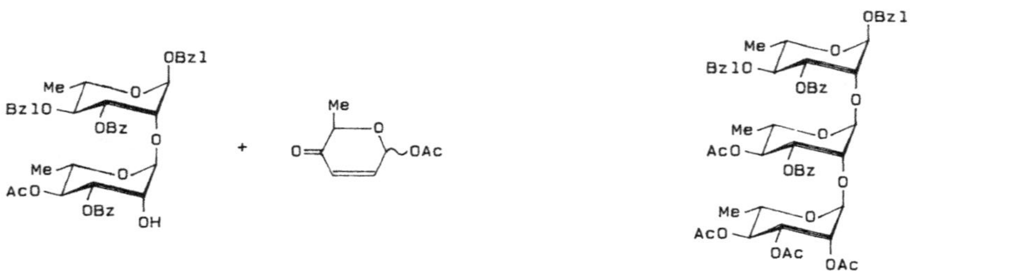

[α]$_D$ −7.4 (chloroform); C-13; Ref.: 283

α-L-Rhap-(1-3)-α-L-Rhap-(1-2)-L-Rha

amorphous; [α]$_D$ -52 (water); C-13; Ref.: 284

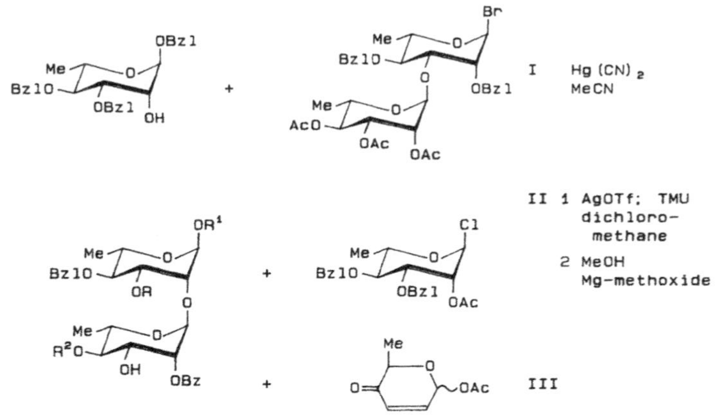

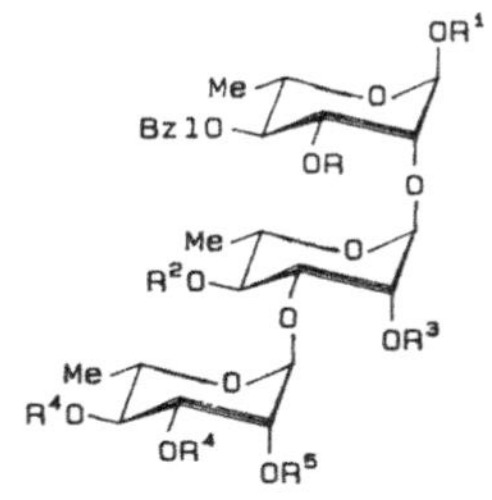

I   R=R^1=Bzl; R^2=R^3=R^4=R^5=Ac
37%; [α]$_D$ -42 (chloroform); C-13; Ref.: 284

II   R=R^4=Bzl; R^1=MCO; R^2=R^3=Bz; R^5=H
69%; [α]$_D$ +24.6 (chloroform); C-13; Ref.: 285

III   R=R^3=Bz; R^1=Bzl; R^2=R^4=R^5=Ac
[α]$_D$ -0 (chloroform); C-13; Ref.: 283

I   1   Hg(CN)$_2$
      MeCN

II   1   AgOTf; TMU
       dichloro-
       methane
    2   MeOH
       Mg-methoxide

III

α-L-Rhap-(1→4)-α-L-Rhap-(1→2)-L-Rha

α-D-(1-OMe): m.p. 88-92°C; $[\alpha]_D$ -70 (water); C-13; Ref.: 286

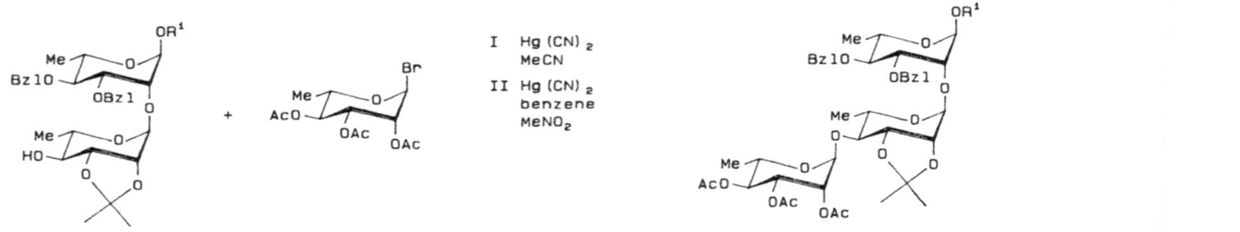

I   Hg(CN)$_2$
    MeCN

II Hg(CN)$_2$
    benzene
    MeNO$_2$

I   R^1=Me: 53%; $[\alpha]_D$ -48 (chloroform); Ref.: 286

II R^1=Bzl: 74.1%; $[\alpha]_D$ -55 (chloroform); C-13; Ref.: 287

α-D-Glcp-(1→3)-α-L-Rhap-(1→3)-L-Rha

α-D-(1-OMe): amorphous powder; $[\alpha]_D$ +15 (MeOH); C-13; Ref.: 288

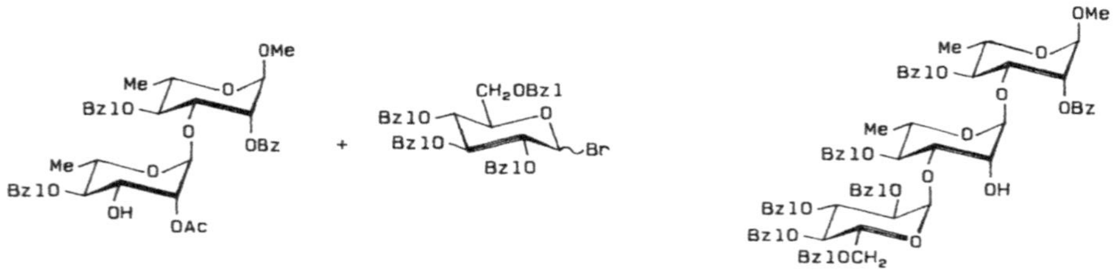

68%; $[\alpha]_D$ +45 (chloroform); C-13; Ref.: 288

β-D-Glcp-(1→3)-α-L-Rhap-(1→3)-L-Rha

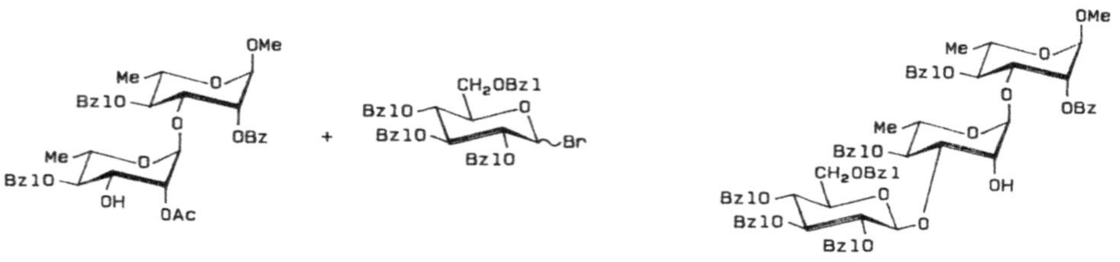

21%; [α]$_D$ +10 (chloroform); C-13; Ref.: 288

α-D-Rhap-(1→2)-α-L-Rhap-(1→3)-L-Rha

m.p. 110—112 °C; [α]$_D$ +2.8 (MeOH); C-13; Ref.: 283

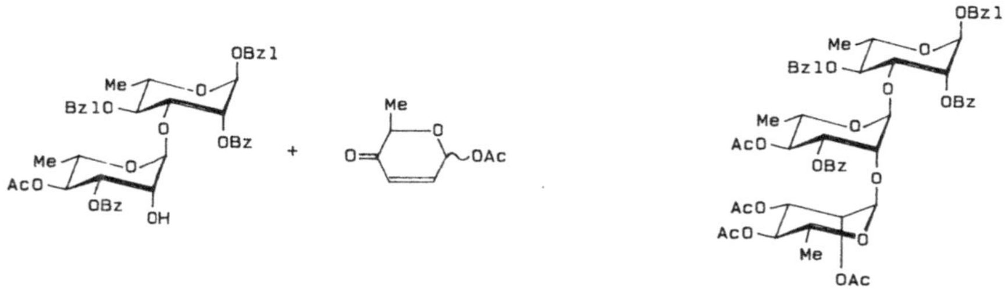

[α]$_D$ +37.4 (chloroform); Ref.: 283

α-L-Rhap-(1→2)-α-L-Rhap-(1→3)-L-Rha

amorphous; [α]$_D$ -40.9 (MeOH); C-13; Ref.: 283

α-p-(1-OMCO); [α]$_{589}$ -77.9 (MeOH); C-13; Ref.: 289

I   1 AgOTf; TMU
        dichloro-
        methane
    2 Mg-methoxide

II AgOTf; TMU
        dichloro-
        methane

III Hg(CN)$_2$; MeCN

I   R=R^4=R^5=Bz1; R^1=MCO; R^2=R^3=Bz; R^6=H
     56%; [α]$_{589}$ +26.6 (chloroform); C-13; Ref.: 281

II   R=R^6=AC; R^1=MCO; R^2=R^3=Bz; R^4=R^5=Bz1
     60%; [α]$_{589}$ +16.1 (chloroform); C-13; Ref.: 289

III R=R^2=R^3=R^4=R^5=Bz; R^1=Me; R^6=Ac
     29%; [α]$_D$ +115.5 (chloroform); C-13; Ref.: 290

IV   R^1=R^3=Bz1; R^2=R^4=Bz; R=R^5=R^6=Ac
     [α]$_D$ -10 (chloroform); Ref.: 283

III 66%; Ref.: 291

III 80%; amorphous: [α]$_D$ +104.8 (chloroform); C-13; Ref.: 292

III 92%; white foam; [α]$_D$ +104.8 (chloroform); C-13; Ref.: 293

α-L-Rhap-(1→3)-α-L-Rhap-(1→3)-L-Rha

amorphous; [α]_D -48 (water); C-13; Ref.: 284

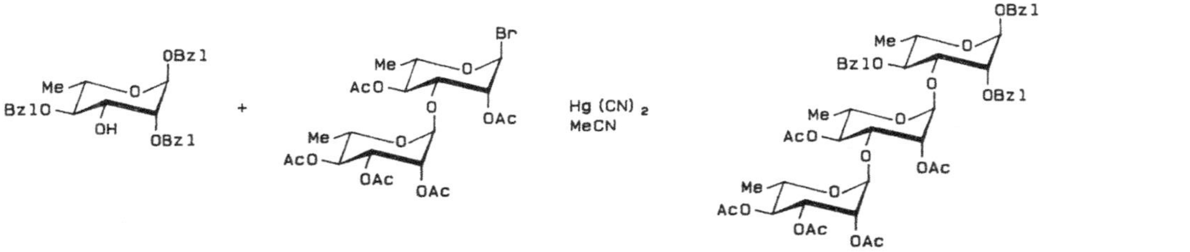

42.6%; [α]_D -37 (chloroform); C-13; Ref.: 284

α-L-Rhap-(1→4)-α-L-Rhap-(1→3)-L-Rha

α-p-(1→OMe); m.p. 92-96°C; [α]_D -86 (water); C-13; Ref.: 286

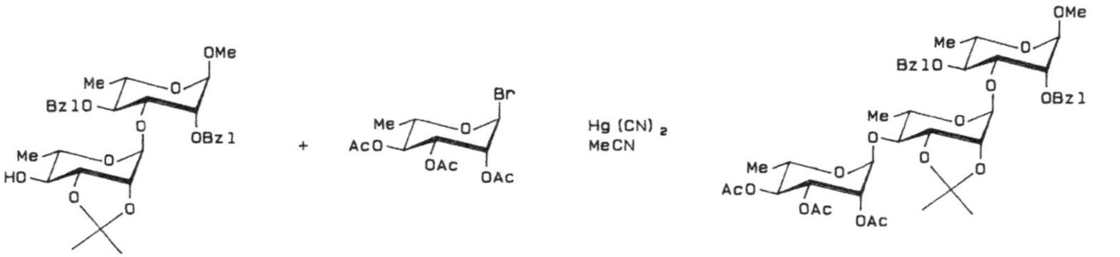

57%; [α]_D -47 (chloroform); Ref.: 286

α-L-Rhap-(1→4)-α-L-Rhap-(1→4)-L-Rha

α-p-(1-OMe); m.p. 148-151 °C; [α]_D -113 (water); C-13; Ref.: 286

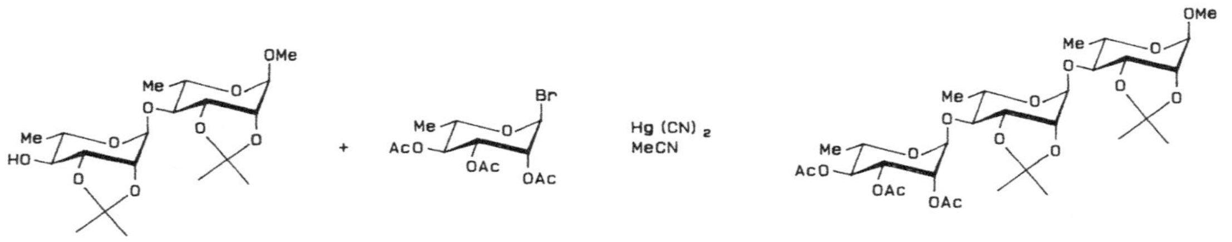

Hg(CN)₂
MeCN

73.3%; m.p. 76-81 °C; [α]_D -72 (chloroform); Ref.: 286

3,6-di-O-Me-β-D-Glcp-(1→4)-2,3-di-O-Me-β-L-Rhap-(1→2)-3-O-Me-L-Rha

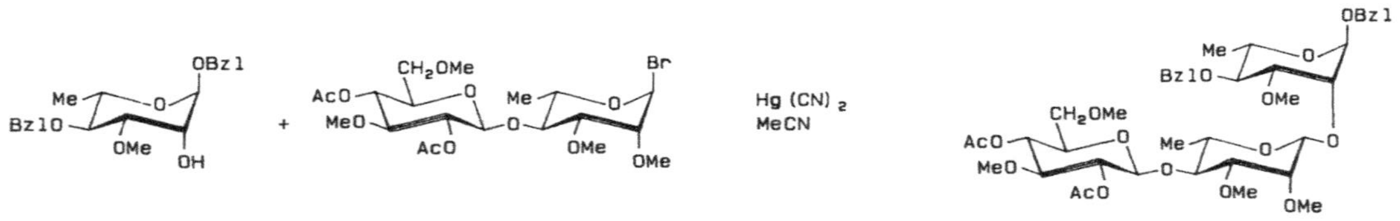

Hg(CN)₂
MeCN

Ref.: 280

α-L-Rhap-(1→4)-β-L-Rhap-(1→2)-L-Rha

[α]$_D$ +2 (MeOH); Ref.: 294

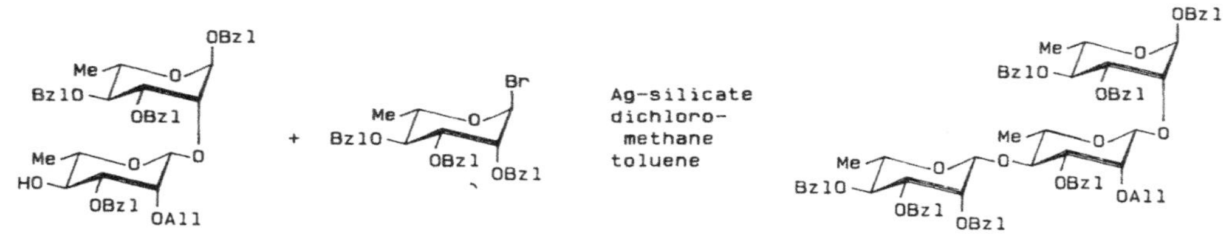

+

Ag-silicate
dichloro-
methane
toluene

66%; [α]$_D$ +17.6 (chloroform); C-13; Ref.: 294

β-D-GalpN-(1→2)-β-L-Rhap-(1→2)-L-Rha

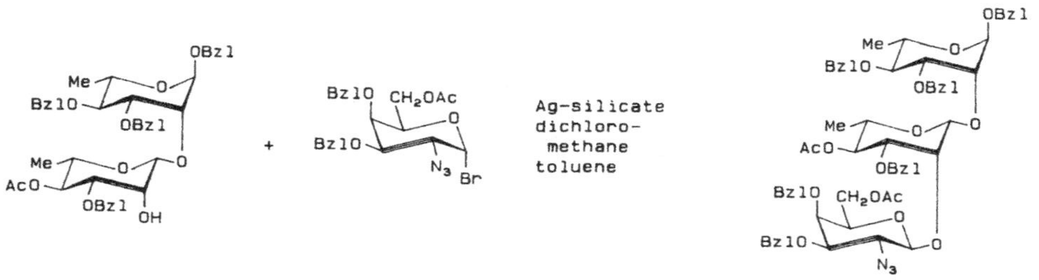

+

Ag-silicate
dichloro-
methane
toluene

62%; [α]$_D$ +16.9 (chloroform); Ref.: 294

β-L-Rhap-(1→4)-β-L-Rhap-(1→4)-L-Rha

[α]$_D$ +16.6 (water); Ref.: 295

Ag-silicate
dichloro-
methane
toluene

72%; m.p. 125°C; [α]$_D$ +1.5 (chloroform); Ref.: 295

α-D-GlcpN-(1→3)-α-D-Galp-(1→4)-L-Rha

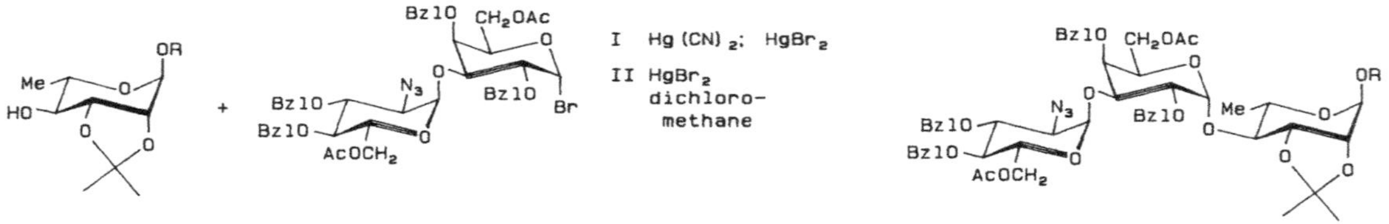

I   Hg(CN)$_2$; HgBr$_2$
II HgBr$_2$ dichloro-
    methane

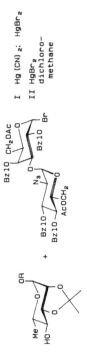

I   R=CH$_2$CCl$_3$: 60%; [α]$_D$ +58 (dichloromethane); Ref.: 296

II R=CH$_2$CCl$_3$: 54%; Ref.: 297

II R=Bzl: 18%; [α]$_D$ 54.7 (dichloromethane); Ref.: 297

β-D-GlcpN-(1→3)-α-D-Galp-(1→4)-L-Rha

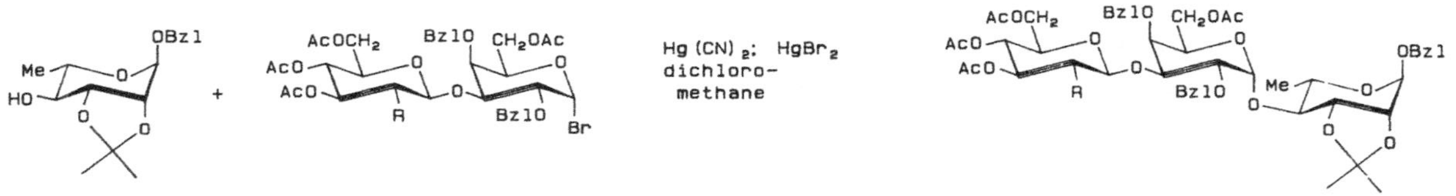

Hg(CN)₂; HgBr₂
dichloro-
methane

R=NHAc; 11%; [α]_D -17.8 (chloroform); Ref.: 297
R=NPhth; 10%; [α]_D +5.6 (chloroform); Ref.: 297

β-D-Manp-(1→4)-α-D-Galp-(1→4)-L-Rha

m.p. 135°C; [α]_D +82.5 (water); Ref.: 298

I   Hg(CN)₂; HgBr₂
II   Hg(CN)₂
III HgBr₂
     dichloro-
     methane

I   R=CH₂CCl₃; 65%; [α]_D +9 (dichloromethane); Ref.: 296
II R=Bzl; 79%; Ref.: 298
III R=Bzl; 77%; [α]_D +15.1 (dichloromethane); Ref.: 298

α-D-GlcpN-(1→3)-β-D-Galp-(1→4)-L-Rha

48%; [α]$_D$ +35.5 (dichloromethane); Ref.: 297

HgBr$_2$
dichloro-
methane

+

β-D-GlcpN-(1→3)-β-D-Galp-(1→4)-L-Rha

Hg(CN)$_2$; HgBr$_2$
dichloro-
methane

+

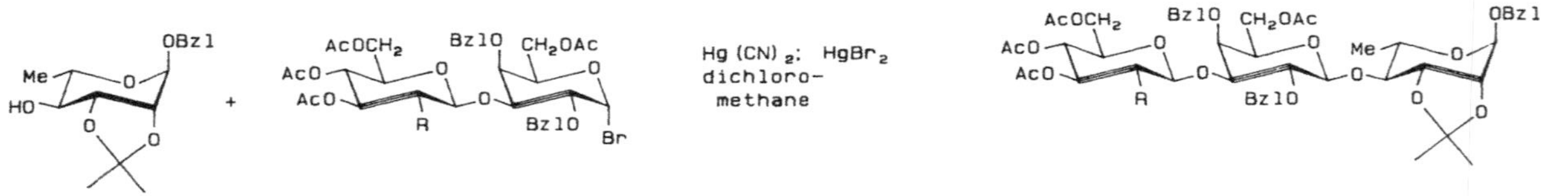

R=NHAc; 57%; [α]$_D$ -45.5 (chloroform); Ref.: 297

R=NPhth; 51%; [α]$_D$ -28.4 (chloroform); Ref.: 297

α-D-GalpN-(1→3)-α-D-GalpN-(1→3)-L-Rha
α-D-GalpNAc-(1→3)-α-D-GalpNAc-(1→3)-L-Rha

α-D-(1→OMCO):   [α]$_D$ +146.5 (water): C-13; Ref.: 285

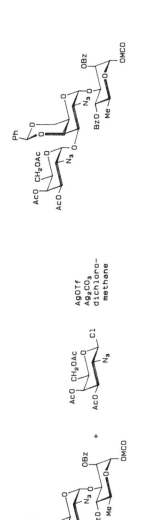

62%: [α]$_D$ +190.5 (chloroform): C-13; Ref.:: 285

α-D-Glcp-(1→2)
                  ⟩ L-Rha
α-L-Rhap-(1→3)

α-D-(1→OMe): m.p. 157-158.5; [α]$_D$ +15.3 (water); Ref.:: 299, 300

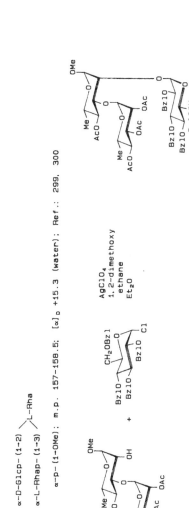

63%: [α]$_D$ +24.4 (chloroform); Ref.:: 299, 300

β-D-GlcpN-(1-2)
\
                    L-Rha
α-D-Glcp-(1-3)
/

β-D-GlcpNAc-(1-2)
\
                    L-Rha
α-D-Glcp-(1-3)
/

α-p-(1-OMe): syrup: $[\alpha]_D$ +45.0 (MeOH); C-13; Ref.: 301

α-p-(1-OMe): $[\alpha]_D$ +43.0 (MeOH); C-13; Ref.: 282

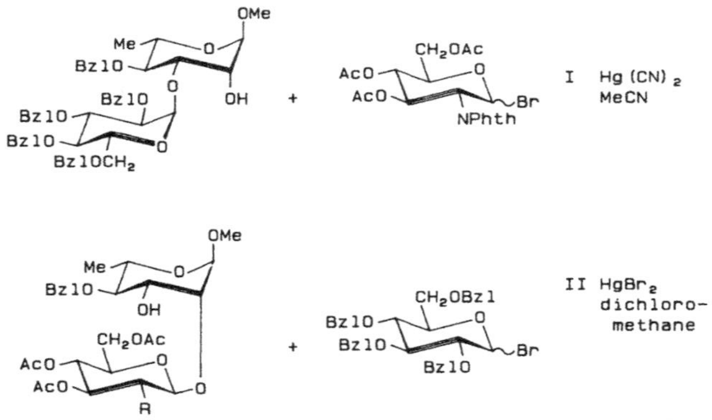

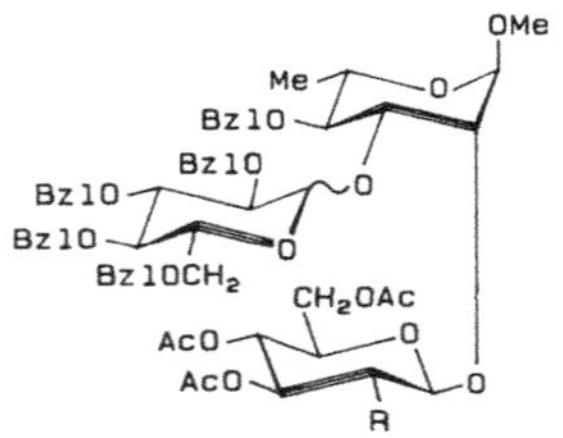

I   Hg(CN)$_2$
    MeCN

II HgBr$_2$
    dichloro-
    methane

I   R=NPhth; β; 80%; $[\alpha]_D$ +73 (chloroform); C-13; Ref.: 301

II R=NHAc; 87%; C-13; Ref.: 282

    R=NPhth; 30%; C-13; Ref.: 282

α-D-Glcp-(1→3)
β-D-ManpN-(1→4) ⟩ L-Rha

α-D-Glcp-(1→3)
β-D-ManpNAc-(1→4) ⟩ L-Rha

[α]_D +27.5 (water); Ref.: 302

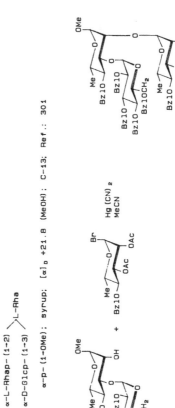

52%; [α]_D +8.3 (chloroform); Ref.: 302

α-L-Rhap-(1→2) ⟩ L-Rha
α-D-Glcp-(1→3)

α-D-(1→OMe); syrup; [α]_D +21.8 (MeOH); C-13; Ref.: 301

71%; [α]_D +5.3 (chloroform); C-13; Ref.: 301

β–D–Glcp–(1→2)
⟩L–Rha
α–L–Rhap–(1→3)

α–ᴅ–(1→OMe): m.p. 156—158 °C; [α]_D –31 (water); Ref.: 299, 300

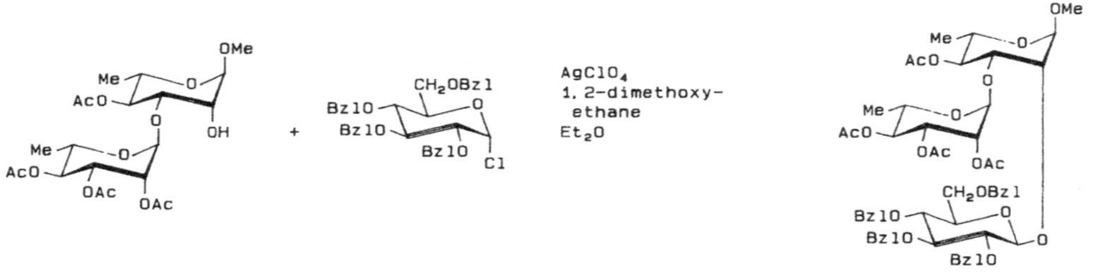

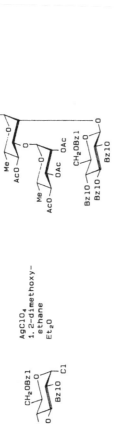

AgClO_4
1.2–dimethoxy–
ethane
Et_2O

14%; m.p. 49—50 °C; [α]_D –13.2 (chloroform); Ref.: 299, 300

β–D–Glcp–(1→2)
⟩L–Rha
α–D–Galp–(1→4)

amorphous: [α]_D +74.5 (water); C–13; Ref.: 303

Hg(CN)_2
benzene
MeNO_2

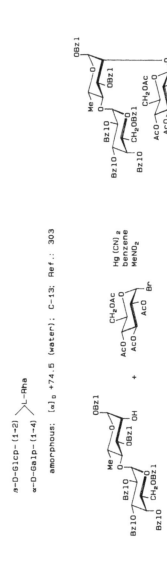

45%; [α]_D +6 (chloroform); C–13; Ref.: 303

β-D-Glcp-(1→3)
β-D-Glcp-(1→4) ⟩L-Rha

α-p-(1-OMe): m.p. 145-147 °C; [α]$_D$ -26.6 (MeOH); C-13; Ref.: 304

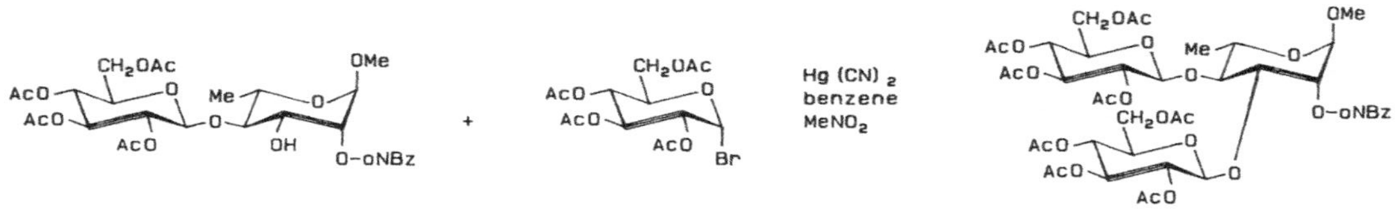

Hg(CN)$_2$
benzene
MeNO$_2$

70%; m.p. 85-87 °C; [α]$_D$ -28.7 (chloroform); C-13; Ref.: 304, 305

β-D-Glcp-(1→3)
β-D-ManpN-(1→4) ⟩L-Rha

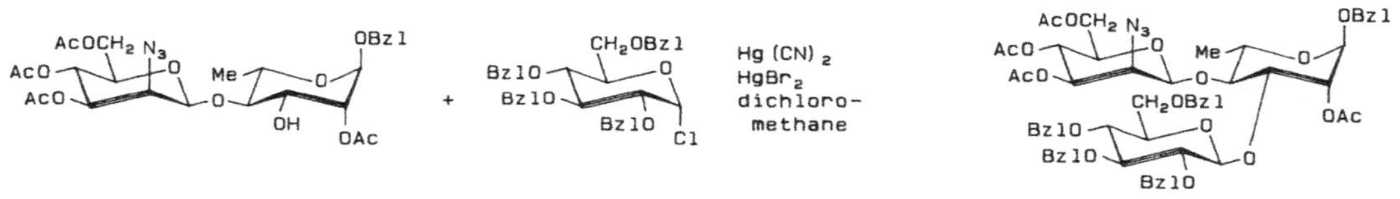

Hg(CN)$_2$
HgBr$_2$
dichloro-
methane

13%; Ref.: 302

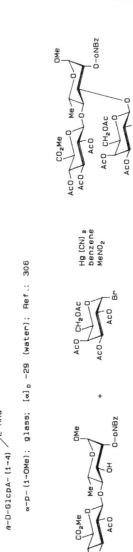

α-D-Galp-(1→3)
β-D-GlcpA-(1→4) ⟩ L-Rha

α-D-(1-OMe): m.p. 155-160 °C; [α]$_D$ +13 (water); Ref.: 306

12%; [α]$_D$ +0.3 (chloroform); C-13; Ref.: 306

β-D-Galp-(1→3)
β-D-GlcpA-(1→4) ⟩ L-Rha

α-D-(1-OMe): glass; [α]$_D$ -29 (water); Ref.: 306

54%; m.p. 103-105 °C; [α]$_D$ -44 (chloroform); C-13; Ref.: 306

α-L-Rhap-(1→3)
                    ⟩ L-Rha
α-L-Rhap-(1→4)

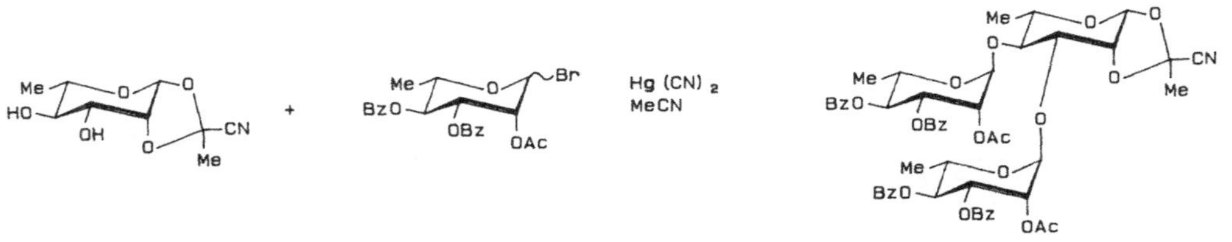

Ref.: 307

β-D-Glcp-(1→4)-α-D-Glcp-(1→6)-D-Gal

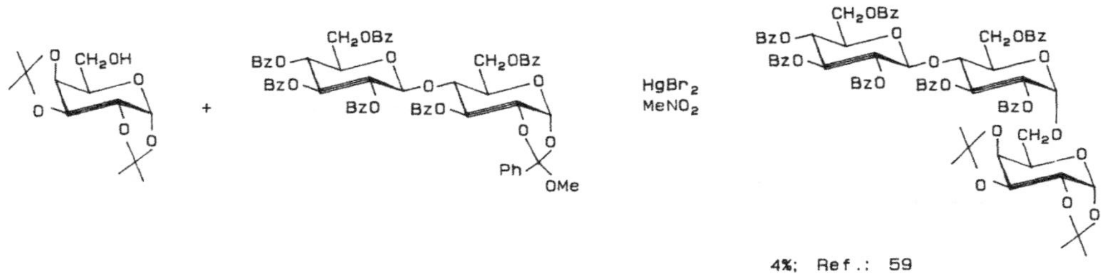

4%; Ref.: 59

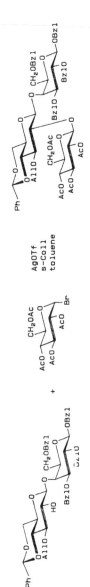

β-D-Glcp-(1→2)-β-D-Glcp-(1→4)-D-Gal

m.p. 248-249°C (dec.): [α]$_D$ +19.7 → +12.7 (water): Ref.: 308

AgOTf
s-Coll
toluene

82%: [α]$_D$ -31.7 (chloroform): C-13: Ref.: 308

β-D-Glcp-(1→4)-β-D-Glcp-(1→6)-D-Gal

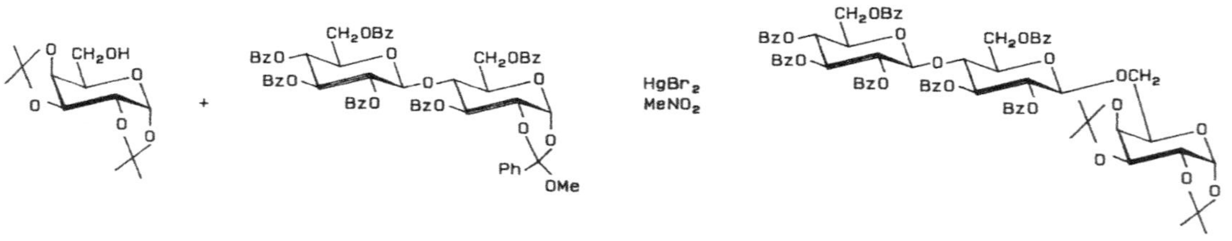

HgBr$_2$
MeNO$_2$

71%: Ref.: 59

β-D-Galp-(1→4)-α-D-GlcpN-(1→6)-D-Gal
β-D-Galp-(1→4)-α-D-GlcpNAc-(1→6)-D-Gal

m.p. 210 °C (dec.): [α]$_D$ +92.9 (water): C-13: Ref.: 309

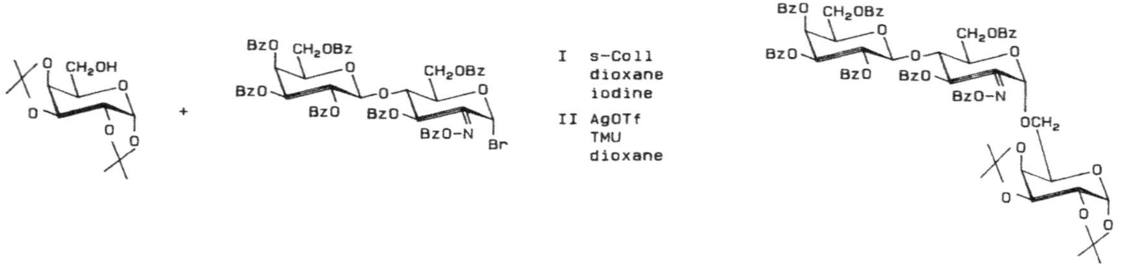

I s-Coll
   dioxane
   iodine

II AgOTf
   TMU
   dioxane

I 78%; m.p. 118-120 °C; [α]$_D$ +60.4 (chloroform): Ref.: 309

II 77%; Ref.: 309

β-D-Galp-(1-3)-β-D-GlcpN-(1-3)-D-Gal
β-D-Galp-(1-3)-β-D-GlcpNAc-(1-3)-D-Gal

m.p. 185-190 °C; $[\alpha]_D$ +19.8 (water): Ref.: 310
m.p. 139-141°C (dec.): $[\alpha]_D$ +13.6 (water): Ref.: 311

β-p-(1-OMe): m.p. 256-258 °C; $[\alpha]_D$ -2.5 (water): C-13: Ref.: 312
β-p-(1-OMCO): C-13: Ref.: 311

III Hg(CN)₂ benzene MeNO₂

II pTSA toluene MeNO₂

III AgOTf s-Coll MeNO₂

I R^1=H; R^2=OMe; R^3=R^4=R^5=Bzl; R^6=NHAc; R, R^1=PhCH
    88%; m.p. 151-153 °C; $[\alpha]_D$ -20.1 (chloroform): Ref.: 312

II R^1=OBzl; R^2=H; R^3=R^4=Bzl; R^5=All; R^6=NHAc; R=Ac
    84%; $[\alpha]_D$ +18.8 (chloroform): Ref.: 310

III R^1=H; R^2=H; R^3=Bz; R^4,R^5=PhCH; R^6=NPhth; R=Ac
    34%; $[\alpha]_D$ +1.6 (chloroform): Ref.: 311

III 24%; m.p. 106-109 °C; $[\alpha]_D$ -15.5 (chloroform): Ref.: 311

β-D-Galp-(1→4)-β-D-GlcpN-(1→3)-D-Gal
β-D-Galp-(1→4)-β-D-GlcpNAc-(1→3)-D-Gal

[α]$_D$ +17.5 (water): Ref.: 311

[α]$_D$ +19.5 (water): Ref.: 313

β-D-(1→OMe): [α]$_D$ +4 (water): C-13: Ref.: 314

β-D-(1→OTCE): [α]$_D$ -4.88 (water): C-13; Ref.: 311

β-D-(1→OMCO): amorphous powder: [α]$_D$ -3.8 (water): Ref.: 311

I

II TMSOTf
    TMU
    dichloro-
    ethane

III AgOTf
    s-Coll
    MeNO$_2$

I   R^1=R^2=R^3=Bzl; R=All; R^4=NHAc
    47%; [α]$_D$ -27 (chloroform): Ref.: 313

II R^1=Me; R^2=R^3=R=Bzl; R^4=NHAc
    18%; [α]$_D$ -14 (chloroform): C-13: Ref.: 314

III R^1=TCE; R^2=Bz; R^3, R=PhCH; R^4=NPhth
    80%; m.p. 252-254°C; [α]$_D$ +8.6 (chloroform): Ref.: 311

    R^1=MCO; R^2=Bz; R^3, R=PhCH; R^4=NPhth
    44%; [α]$_D$ +31.5 (chloroform): Ref.: 311

β-D-Glcp-(1→4)-β-D-GlcpN-(1→6)-D-Gal

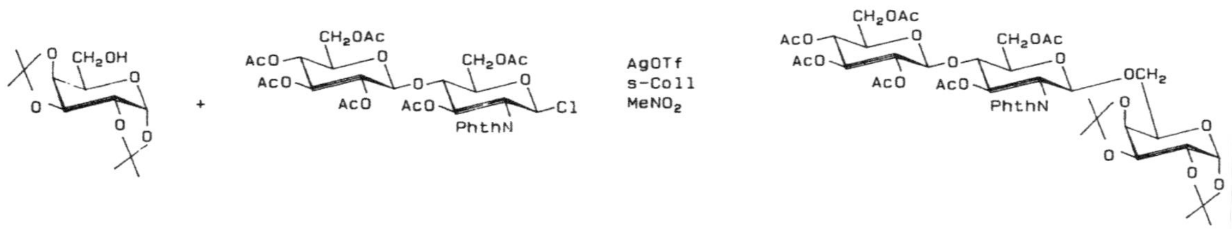

AgOTf
s-Coll
MeNO$_2$

65%; m.p.: 235–237 °C; [α]$_D$ –20.7 (chloroform); Ref.: 191

β-D-Galp-(1→3)-β-D-GlcpN-(1→6)-D-Gal
β-D-Galp-(1→3)-β-D-GlcpNAc-(1→6)-D-Gal

m.p. 177-184 °C; $[\alpha]_D$ -9.5 → -1.2 (water); Ref.: 313

m.p. 175-178 °C; $[\alpha]_D$ -0.9 (water); Ref.: 311

β-D-(1-O-p-Nitrophenyl); amorphous; $[\alpha]_D$ -53 (water); Ref.: 315

β-D-(1-OMCO); m.p. 185-188 °C; $[\alpha]_D$ -20.9 (water); Ref.: 311

I

II pTSA
    benzene
    MeNO$_2$

III AgOTf
    s-Coll
    MeNO$_2$

I R^1=R^2=R^4=Bzl; R^3=All
   79%; $[\alpha]_D$ -4 (chloroform); Ref.: 313

II R^1=p-Nitrophenyl; R^2=R^3=Ac; R^4=H
   19.5%; m.p. 140-142 °C; $[\alpha]_D$ -6.5 (chloroform); Ref.: 315

III R^1=H; R,R^2=OC(Me)$_2$; m.p. 119-121 °C; $[\alpha]_D$ -39.1 (chloroform); Ref.: 311
    76%
    R=H; R^1=OMCO; R^2=Bz
    71%; m.p. 87-91 °C; Ref.: 311

β-D-Galp-(1→4)-β-D-GlcpN-(1→6)-D-Gal
β-D-Galp-(1→4)-β-D-GlcpNAc-(1→6)-D-Gal

[α]_D +4 (water): Ref.: 313

m.p. 155-159 °C; [α]_D +9.6 (water): Ref.: 311

β-p-(1-OMCO): m.p. 202-205 °C; [α]_D -15.7 (water): C-13: Ref.: 311

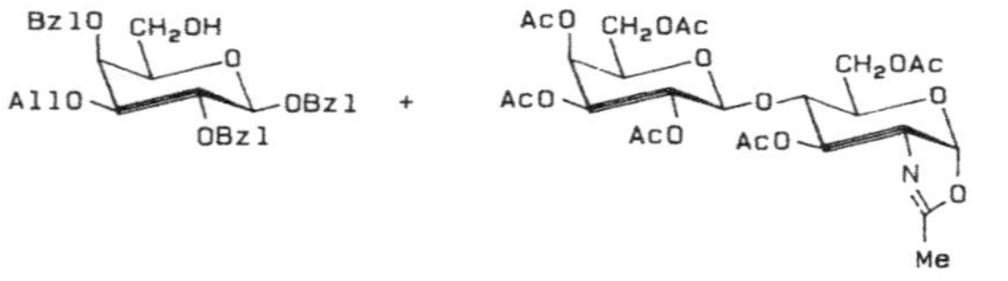

I

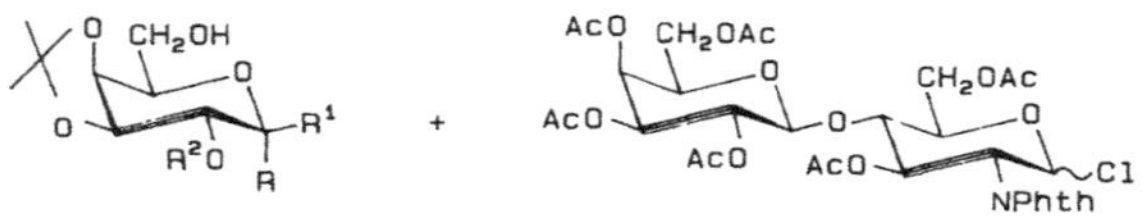

II   AgOTf
s-Coll
MeNO₂

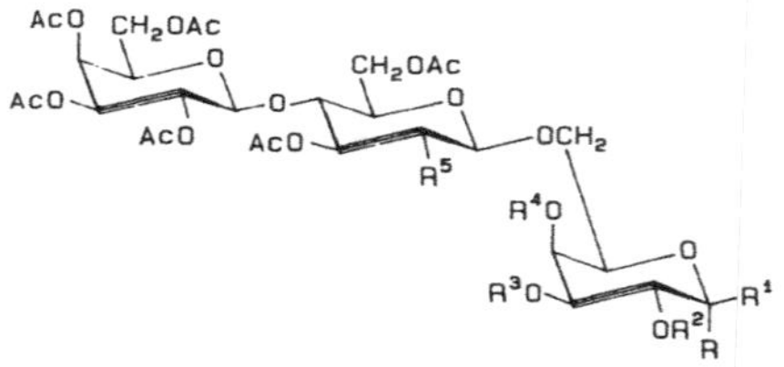

I   R¹=OBzl; R²=R⁴=Bzl; R³=All; R⁵=NHAc; R=H
78%; [α]_D -27 (chloroform); Ref.: 313

II R¹=H; R,R²=OC(Me)₂; R³,R⁴=C(Me)₂; R⁵=NPhth; R=H
80%; m.p. 201-202°C; [α]_D -19.6 (chloroform); Ref.: 311

R¹=OMCO; R²=Bz; R³,R⁴=C(Me)₂; R⁵=NPhth; R=H
64%; [α]_D +21.3 (chloroform); Ref.: 311

β-D-Fucp-(1→4)-β-D-GlcpN-(1→6)-D-Gal

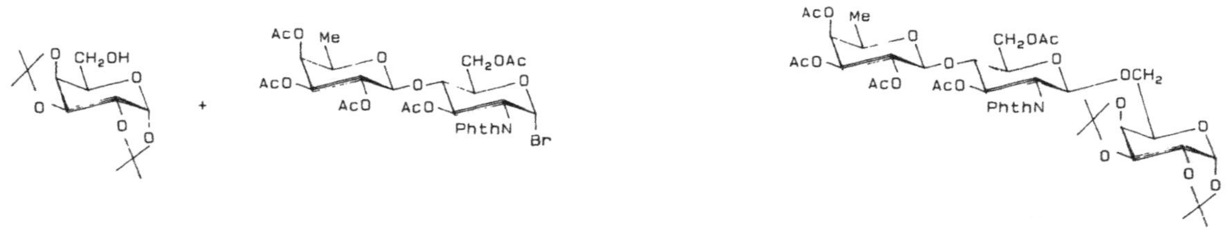

54%; Ref.: 191

α-D-Manp-(1→2)-α-D-Manp-(1→3)-D-Gal

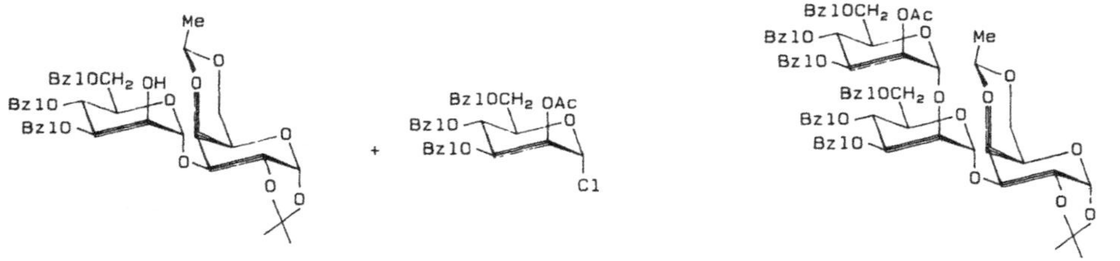

62%; C-13; Ref.: 316

α-D-Manp-(1→2)-β-D-Manp-(1→3)-D-Gal

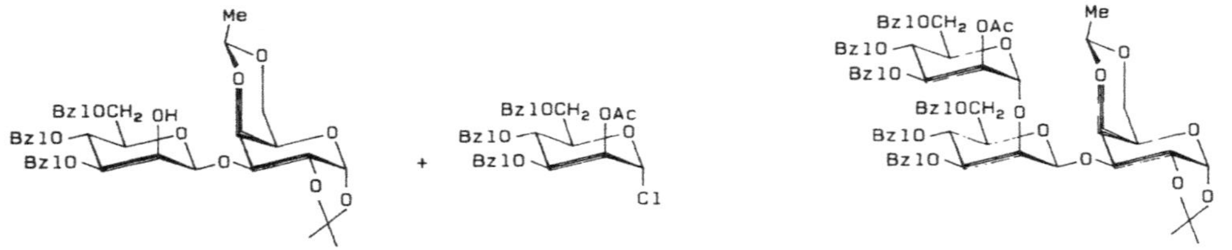

65%: C-13: Ref.: 316

β-D-Galp-(1→4)-β-D-Manp-(1→6)-D-Gal

m.p. 210 °C (dec.); [α]$_D$ +17.7 (DMSO); C-13; Ref.: 309

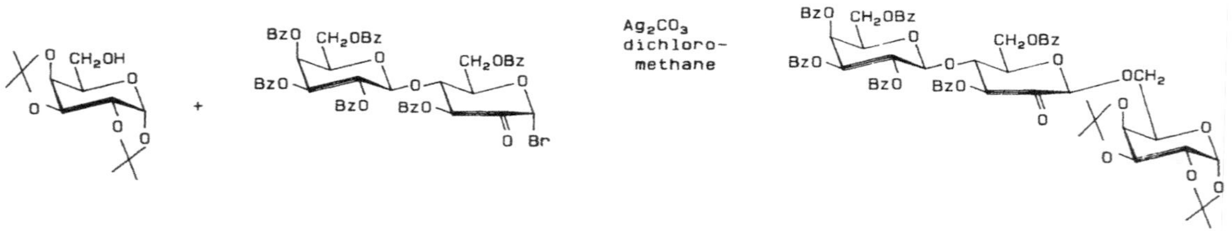

Ag$_2$CO$_3$
dichloro-
methane

62%: [α]$_D$ +38.4 (chloroform); Ref.: 309

β-D-Galp-(1→4)-β-D-ManpN-(1→6)-D-Gal
β-D-Galp-(1→4)-β-D-ManpNAc-(1→6)-D-Gal

m.p. 162 °C (dec.); $[\alpha]_D$ +11 (water); Ref.: 309

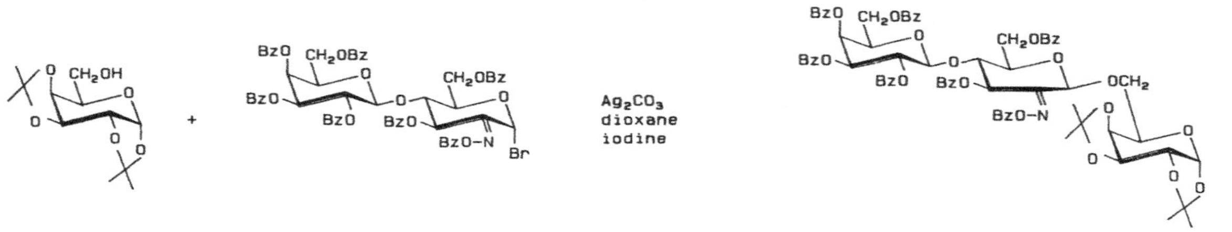
CH₂OH
+
BzO
CH₂OBz
BzO
BzO-N
Br
Ag₂CO₃
dioxane
iodine
OCH₂

81%; m.p. 108–110 °C; $[\alpha]_D$ +8.1 (chloroform); Ref.: 309

β-D-Glcp-(1→4)-α-L-Rhap-(1→3)-D-Gal

$[\alpha]_D$ -19 (MeOH): Ref.: 317

$[\alpha]_D$ -0.1 (water): -14.0 (MeOH): Ref.: 318

β-p-(1-O-p-Nitrophenyl): m.p. 228-230 °C; $[\alpha]_D$ -76.7 (water): Ref.: 319

I   Hg(CN)₂
     MeCN

II   1 Hg(CN)₂
       MeCN
     2 NaOMe

III Hg(CN)₂
     MeCN
     MeNO₂
     s-Coll

I   65%: $[\alpha]_D$ -43 (chloroform): Ref.: 320

II   37%: m.p. 228-230 °C:
     $[\alpha]_D$ -76.7 (water): Ref.: 319

I   R=Ac: 55%: m.p. 104-108 °C:
     $[\alpha]_D$ -25 (chloroform): Ref.: 317

III R=Ac: 70%: $[\alpha]_D$ -17 (chloroform): Ref.: 318

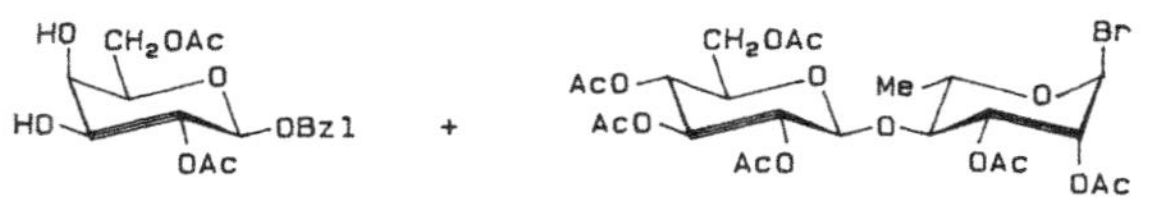

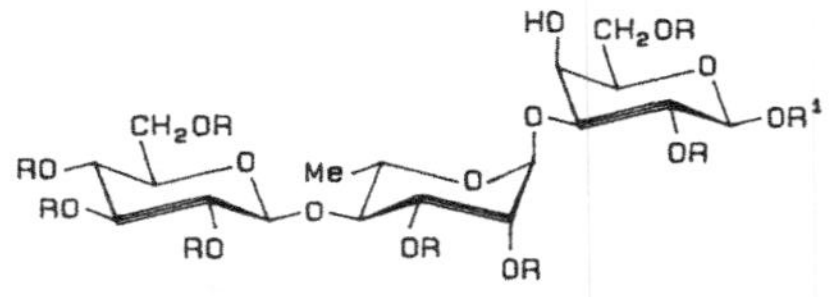

IV R=Ac; R¹=Bzl: 15%
    $[\alpha]_D$ -33 (chloroform): Ref.: 321

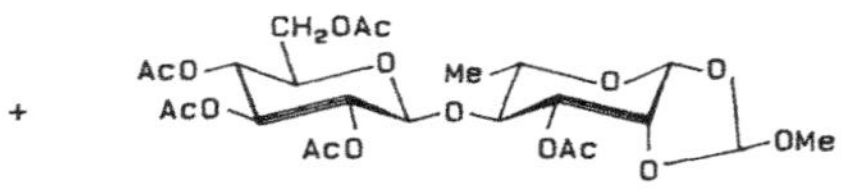

V   R=Ac; R¹=Bzl: 15%; Ref.: 321

I   R,R=C(Me)₂: 81%;
    $[\alpha]_D$ +10.4 (chloroform): Ref.: 318

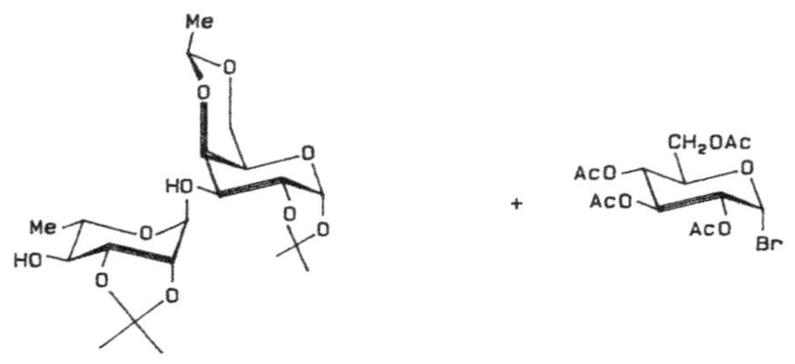

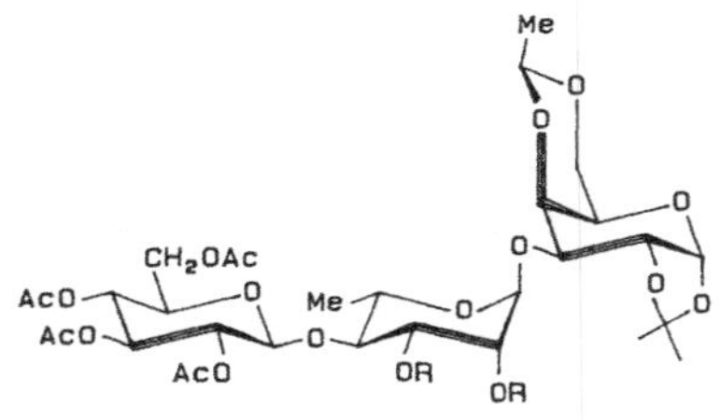

IV Ag₂CO₃
    benzene

V   HgBr₂
    MeNO₂

I

α-D-Manp-(1→4)-α-L-Rhap-(1→3)-D-Gal

syrup; $[\alpha]_D$ +22 (water); Ref.: 322

$[\alpha]_D$ +27.2 (water); Ref.: 318

$[\alpha]_D$ +27 (water); C-13; Ref.: 323

α-p-(1-O-p-Trifluoroacetamidophenyl); amorphous powder; $[\alpha]_D$ +106 (water); C-13; Ref.: 324

I 15%; $[\alpha]_D$ −15.3 (chloroform); Ref.: 322

II 100%; $[\alpha]_D$ −28 (chloroform); Ref.: 323

II 77%; $[\alpha]_D$ +20.4 (chloroform); Ref.: 318

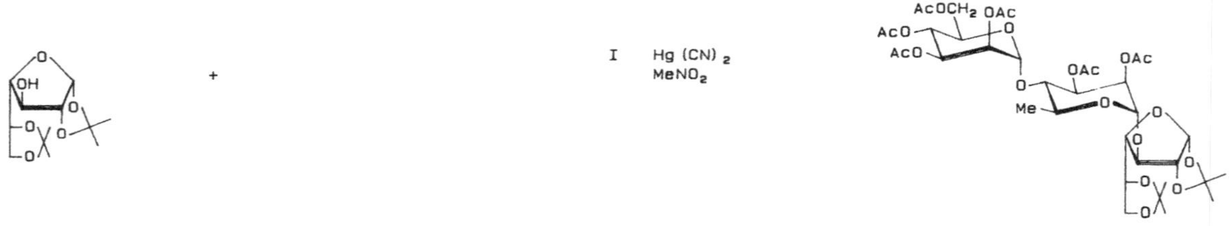

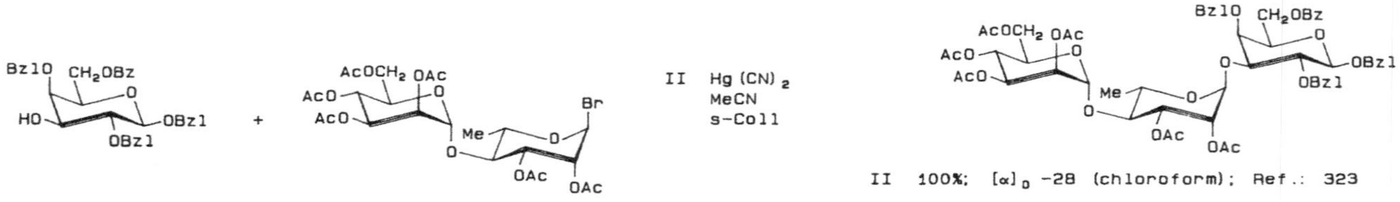

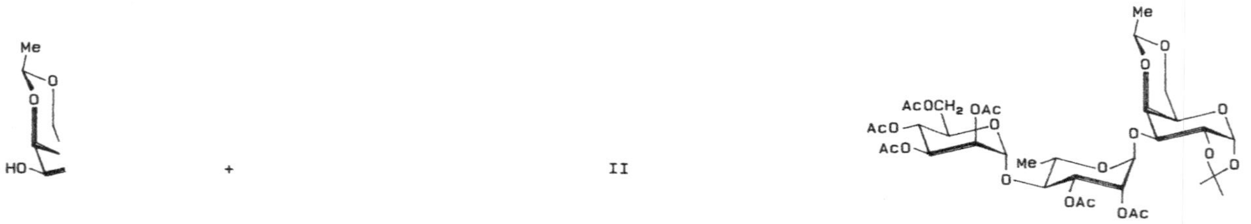

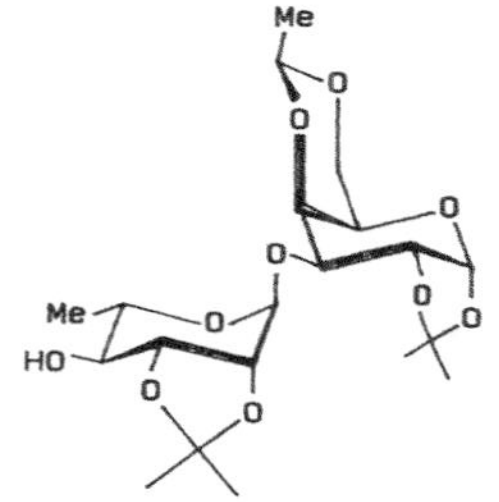

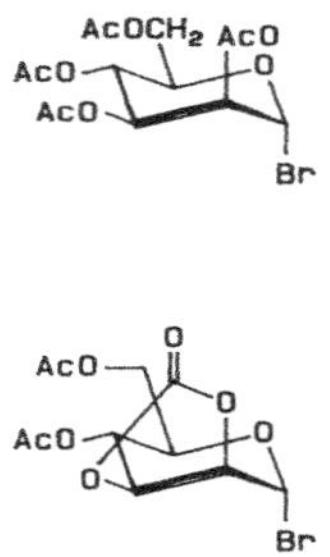

III Hg(CN)₂
     MeCN

III R=R⁵=AC; R⁴,R⁴=C(Me)₂ [α]_D +36.4 (chloroform): Ref.: 318
     72%; m.p. 169-171°C;

IV   Ag₂O
     AgClO₄
     dichloro-
     methane

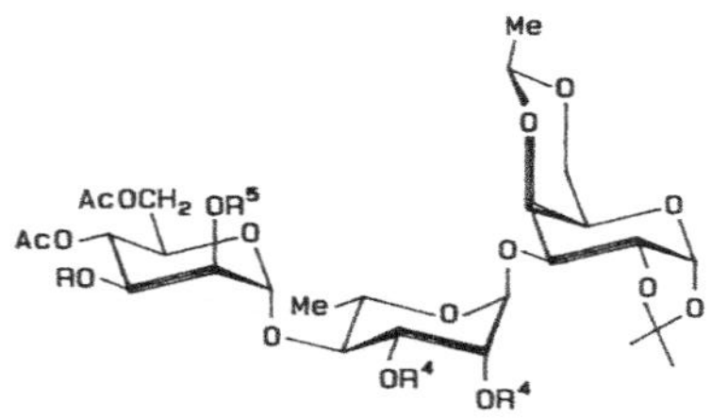

IV   R,R⁵= =CO; R⁴,R⁴=C(Me)₂ [α]_D +16 (chloroform): Ref.: 318
     4.4%;

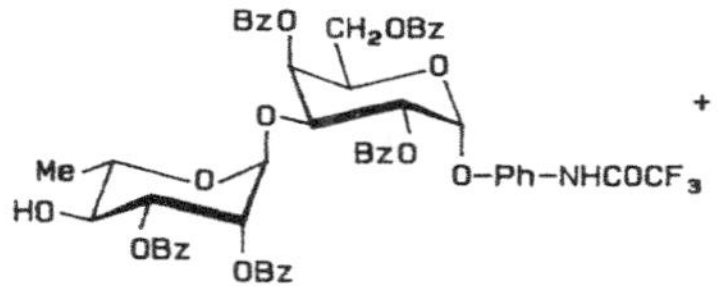

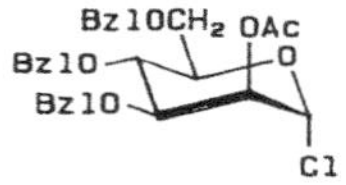

V   AgOTf
     toluene
     MeNO₂

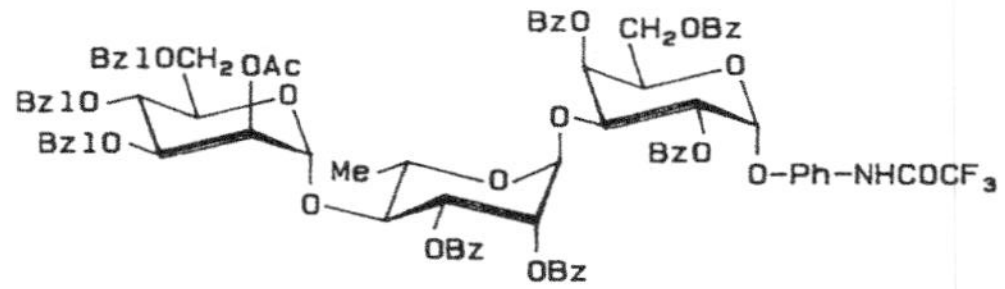

V   73%; m.p. 97-98°C; [α]_D +112 (chloroform); C-13: Ref.: 324

β-D-Manp-(1-4)-α-L-Rhap-(1-3)-D-Gal

$[α]_D$ -24 (water): Ref.: 325

$[α]_D$ -13 (water): Ref.: 326

$[α]_D$ -15.2 (water): Ref.: 318

$[α]_D$ -15 (water): Ref.: 327

β-p-(1-OAll): m.p. 232-235°C; $[α]_D$ -51.5 (water): C-13: Ref.: 328, 329, 330

α-p-(1-O-p-Trifluoroacetamidophenyl): amorphous powder: $[α]_D$ +57 (water): C-13: Ref.: 324

α-p-(1-O-p-Trifluoroacetamidophenyl): m.p. 260°C (dec.): Ref.: 276

β-p-(1-O-p-Nitrophenyl): m.p. 218-220°C; $[α]_D$ -82 (MeOH): Ref.: 331

β-p-(1-O-p-Aminophenyl): m.p. 165-172°C; $[α]_D$ -57 (water); Ref.: 331

I   Ag$_2$O
    AgClO$_4$
    dichloro-
    methane

II AgOTf
    toluene
    MeNO$_2$

I   66%; $[α]_D$ -11.8 (chloroform): Ref.: 318

II 2%; m.p. 105-109°C; $[α]_D$ +55 (chloroform): Ref.: 324

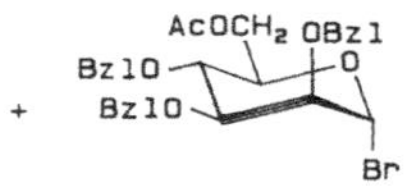

III 1 Ag-zeolite
     dichloro-
     methane
  2 PdC; H₂
  3 NaOMe

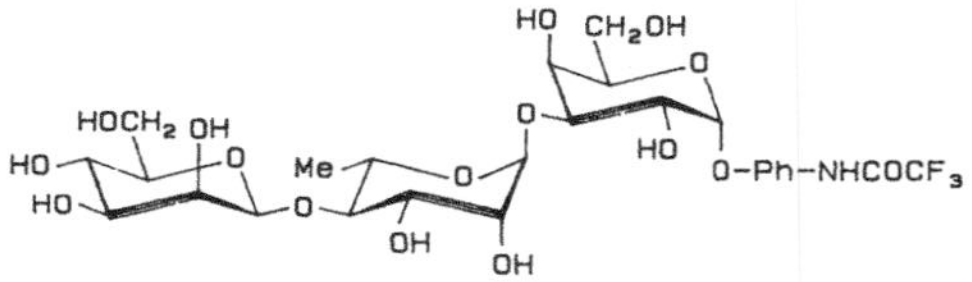

III 31%: m.p. 260 °C (dec.); Ref.: 276

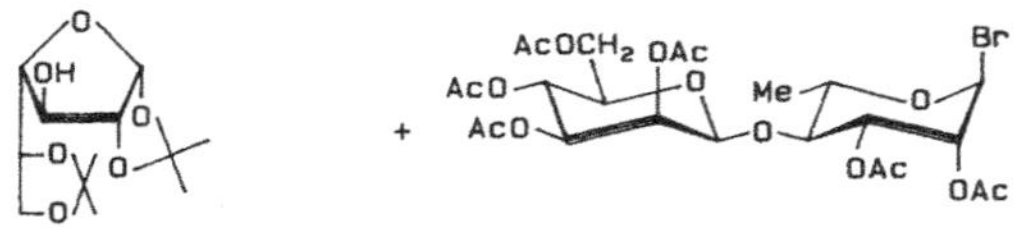

IV Hg(CN)₂
   MeCN

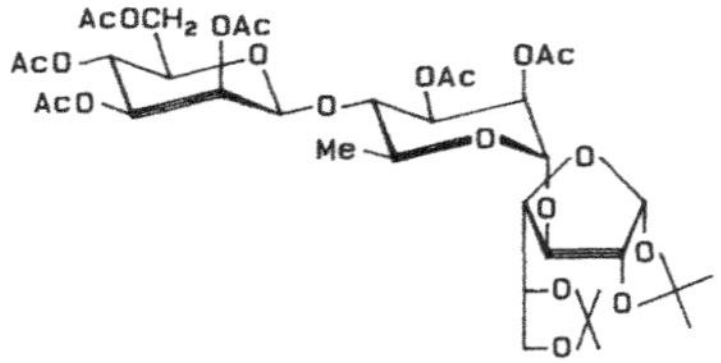

IV 22%; [α]ᴅ -41.5 (chloroform); Ref.: 325

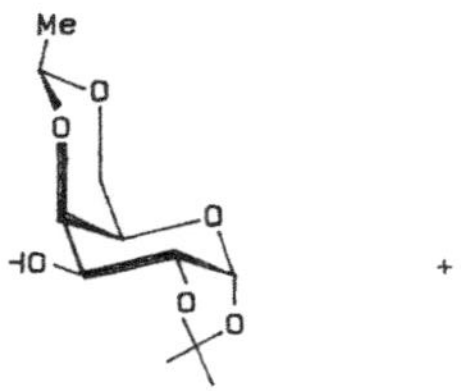

IV

V Hg(CN)₂
  MeCN
  s-Coll

VI Hg(CN)₂
   MeCN
   MeNO₂
   s-Coll

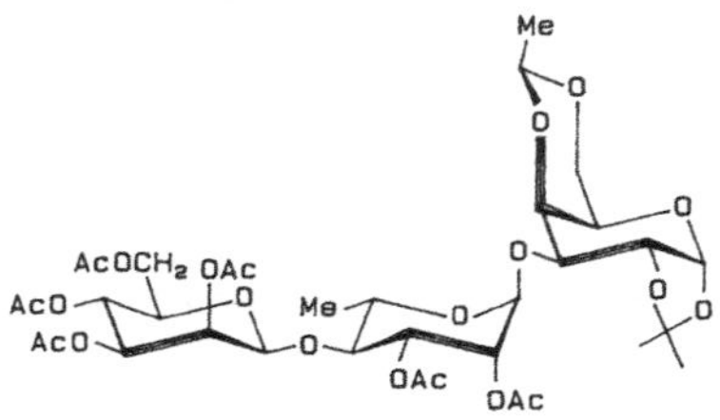

IV 48%; m.p. 87-92 °C; [α]ᴅ -23 (chloroform); Ref.: 326

VI 75%; [α]ᴅ -27.2 (chloroform); Ref.: 318

IV 48%; m.p. 221-222 °C;
      [α]ᴅ -28 (chloroform); Ref.: 332

V 64%; m.p. 221-222 °C; Ref.: 332

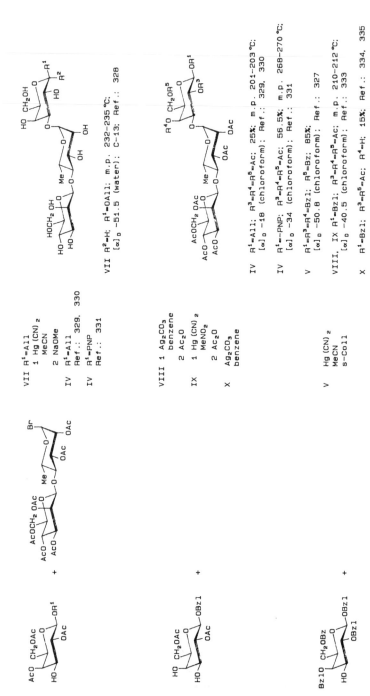

VII R^1=All
  1 Hg (CN)$_2$
   MeCN
  2 NaOMe

IV R^1=All
  Ref.: 329, 330

IV R^1=PNP
  Ref.: 331

VIII 1 Ag$_2$CO$_3$
   benzene
  2 Ac$_2$O

IX 1 Hg (CN)$_2$
   MeNO$_2$
  2 Ac$_2$O

X   Ag$_2$CO$_3$
  benzene

V   Hg (CN)$_2$
  MeCN
  s-Coll

VII R^2=H; R^1=OAll: m.p. 232—235 °C;
    [α]$_D$ −51.5 (water); C-13; Ref.: 328

IV R^1=All; R^3=R^4=R^5=Ac: 25%; m.p. 201—203 °C;
    [α]$_D$ −18 (chloroform): Ref.: 329, 330

IV R^1=-PNP: R^3=R^4=R^5=Ac: 56.5%; m.p. 268—270 °C;
    [α]$_D$ −34 (chloroform): Ref.: 331

V   R^1=R^3=R^4=Bzl; R^5=Bz: 85%;
    [α]$_D$ −50.8 (chloroform): Ref.: 327

VIII, IX R^1=Bzl; R^3=R^4=R^5=Ac: m.p. 210—212 °C;
    [α]$_D$ −40.5 (chloroform): Ref.: 333

X   R^1=Bzl; R^3=R^5=Ac; R^4=H: 15%; Ref.: 334, 335

β-D-Glcp-(1→4)-α-L-Rhap-(1→4)-D-Gal

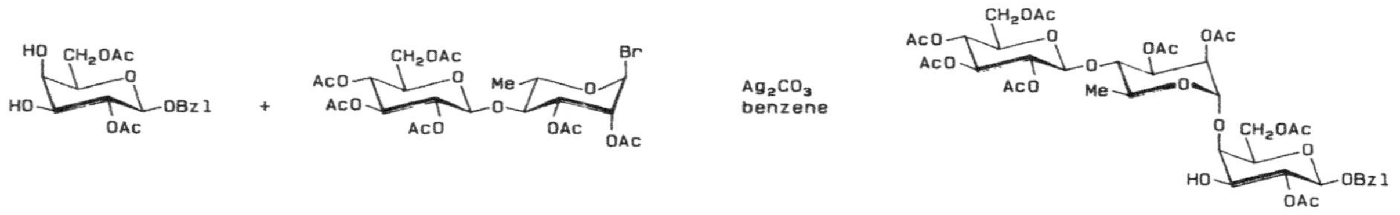

15%; m.p. 204-206 °C; $[\alpha]_D$ -61.8 (chloroform); Ref.:: 321

β-D-Manp-(1→4)-α-L-Rhap-(1→4)-D-Gal

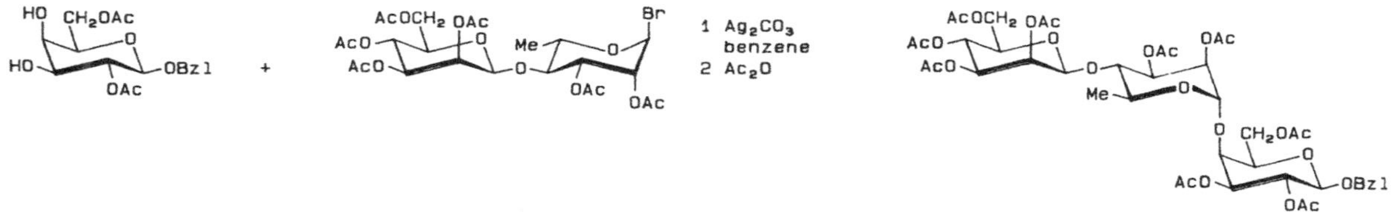

$[\alpha]_D$ -33 (chloroform); Ref.:: 333

β-D-Glcp-(1→4)-α-L-Rhap-(1→6)-D-Gal

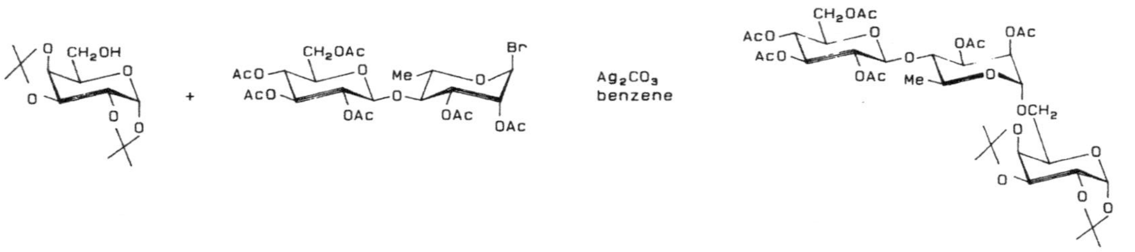

27%; [α]$_D$ -57 (chloroform); Ref.: 321

α-L-Rhap-(1-2)-α-L-Rhap-(1-6)-D-Gal

amorphous; [α]$_D$ -53 (water); C-13; Ref.: 336

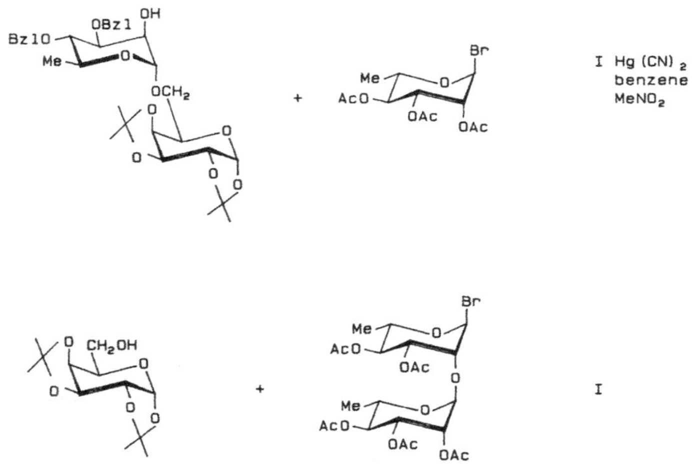

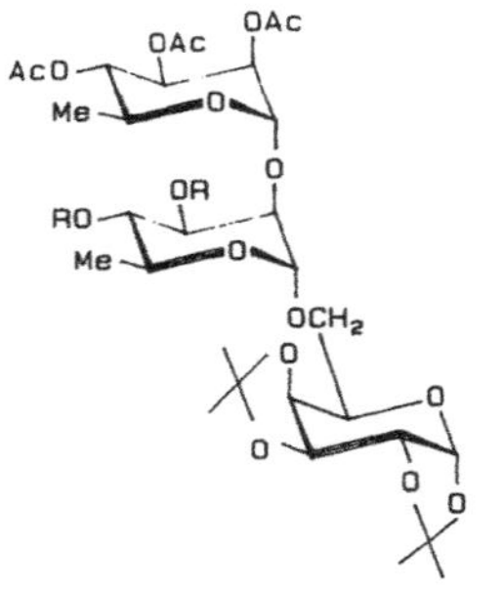

I Hg(CN)$_2$
benzene
MeNO$_2$

R=Bzl: 52.7%; [α]$_D$ -63.3 (chloroform); Ref.: 336

R=Ac: 47.2%; [α]$_D$ -53.5 (chloroform); C-13; Ref.: 336

α-L-Rhap-(1→3)-α-L-Rhap-(1→6)-D-Gal

m.p. 215°C; [α]₀ -43 (water); Ref.: 82

[α]₀ -35.5 (water); C-13; Ref.: 284

m.p. 215°C; [α]₀ -46.1 (water); C-13; Ref.: 336

I   1 Hg(CN)₂
     2 H⁺
     3 Ac₂O

II   Hg(CN)₂
     benzene
     MeNO₂

III Hg(CN)₂
     MeCN

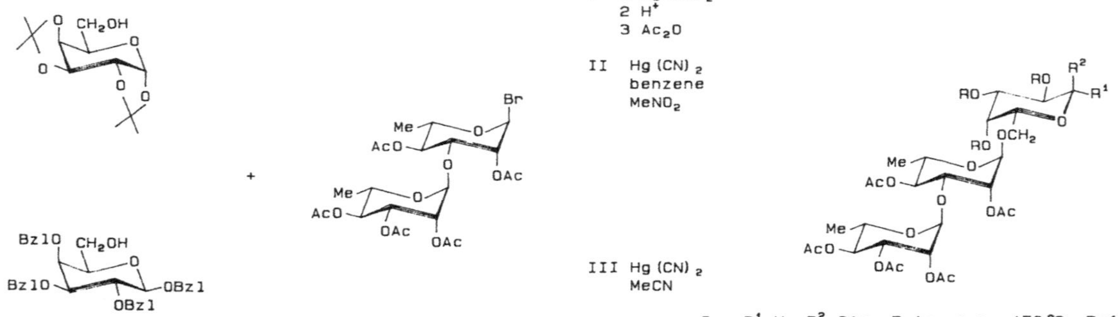

I   R¹=H; R²=OAc; R=Ac; m.p. 152°C; Ref.: 82

II   R¹=H; R², R=OC(Me)₂; R, R=C(Me)₂
     65.5%; [α]₀ -57.4 (chloroform); C-13; Ref.: 336

III R¹=OBzl; R²=H; R=Bzl; [α]₀ -42.3 (chloroform); Ref.: 284

α-L-Rhap-(1→4)-α-L-Rhap-(1→6)-D-Gal

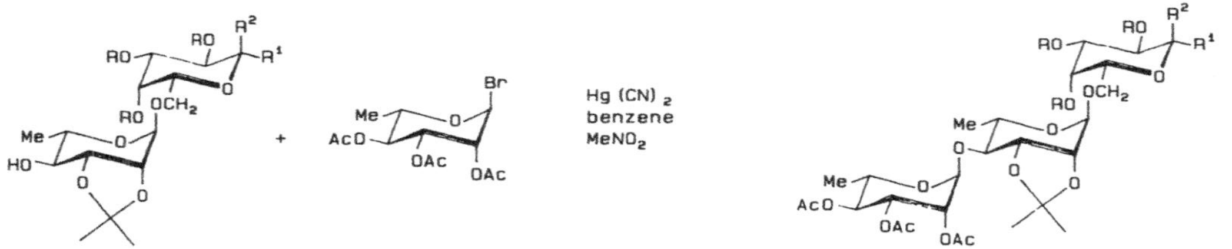

Hg(CN)₂
benzene
MeNO₂

R=Bzl; R¹=OBzl; R²=H; 79.6%; m.p. 126—128 °C; [α]$_D$ −79 (chloroform); Ref.: 337

R¹=H; R², R=OC(Me)₂; R, R=C(Me)₂; 61.5%; [α]$_D$ −94.5 (chloroform); Ref.: 338

α-D-Manp-(1→4)-β-L-Rhap-(1→3)-D-Gal

[α]$_D$ +82 (water); C-13; Ref.: 323

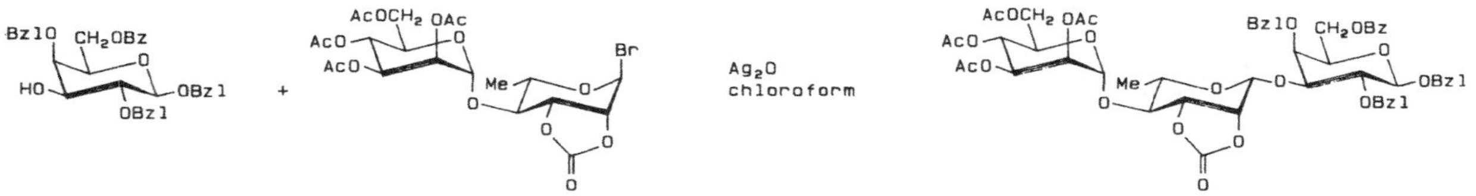

Ag₂O
chloroform

44%; [α]$_D$ +29.5 (chloroform); Ref.: 323

β-D-Manp-(1→4)-β-L-Rhap-(1→3)-D-Gal

[α]$_D$ +40 (water); C-13; Ref.: 332

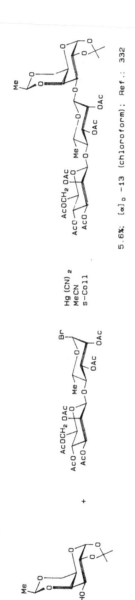

Hg (CN)$_2$
MeCN
s-Coll

5.6%; [α]$_D$ -13 (chloroform); Ref.: 332

α-D-Galp-(1→6)-α-D-Galp-(1→6)-D-Gal

β-D- : solid glass; [α]$_D$ +117 (water); Ref.: 339

Et$_4$NCl
Et$_3$N
dichloro-
ethane

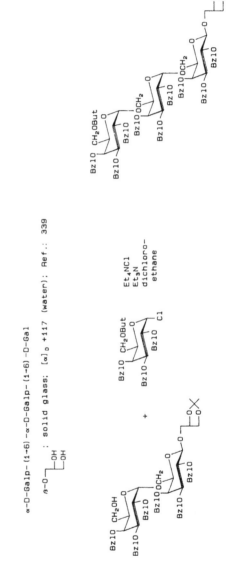

β-D-Galp-(1→2)-β-D-Galp-(1→2)-D-Gal

β-D-(1→OMe); [α]_D +19.3 (water); Ref.: 340

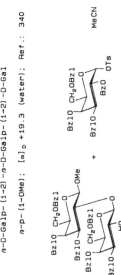

[α]_D +2.2 (chloroform); C-13; Ref.: 340

α-D-Galp-(1→3)-β-D-Galp-(1→3)-D-Gal

33.5%; [α]_D +96 (chloroform); C-13; Ref.: 341

β-D-Galp-(1-3)-β-D-Galp-(1-3)-D-Gal

m.p. 210-220 °C (dec.); [α]$_D$ +52 (water); C-13; Ref.: 342

β-D-(1-OMe); m.p. 233-235 °C; [α]$_D$ +31.4 (water); C-13; Ref.: 341

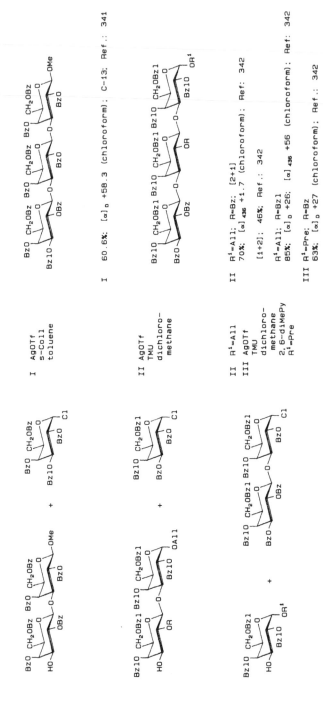

I   AgOTf
     s-Coll
     toluene

I   60.6%; [α]$_D$ +58.3 (chloroform); C-13; Ref.: 341

II   AgOTf
     TMU
     dichloro-
     methane

II   R^1=All
III AgOTf
     TMU
     dichloro-
     methane
     2,6-diMePy
     R^1=Pre

II   R^1=All; R=Bz; [2+1]
70%; [α]$_{436}$ +1.7 (chloroform); Ref.: 342

[1+2]; 46%; Ref.: 342

R^1=All; R=Bzl
85%; [α]$_D$ +26; [α]$_{436}$ +56 (chloroform); Ref: 342

III R^1=Pre; R=Bz
63%; [α]$_D$ +27 (chloroform); Ref.: 342

β-D-Galp-(1→4)-β-D-Galp-(1→4)-D-Gal

α-p-(1-OPr); white solid; C-13; Ref.: 343

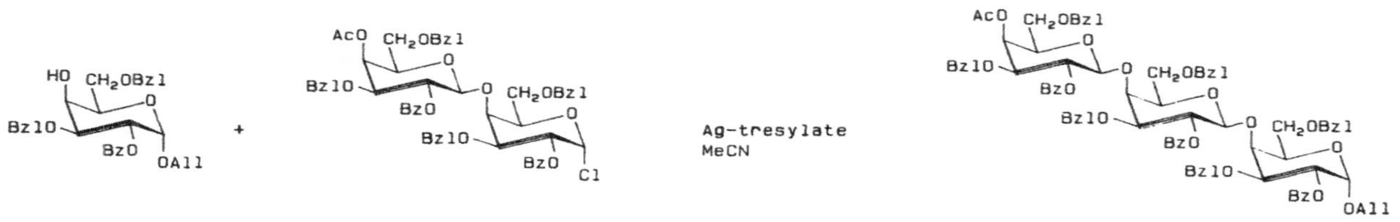

+

Ag-tresylate
MeCN

85%; m.p. 126-127°C; [α]_D +70.5 (chloroform); C-13; Ref.: 343

β-D-Glcp-(1→2)-β-D-Galp-(1→6)-D-Gal

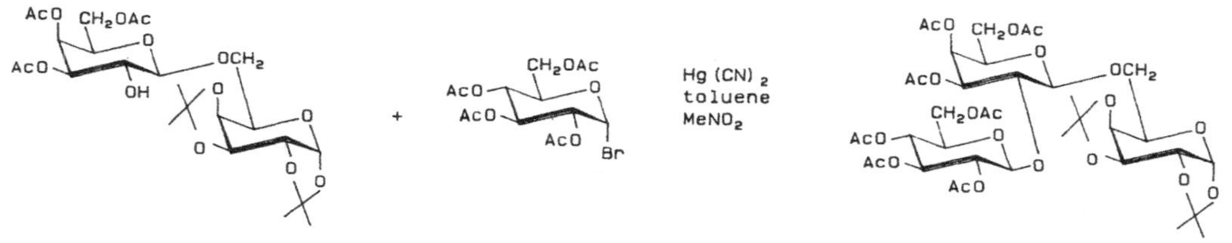

+

Hg(CN)₂
toluene
MeNO₂

68%; m.p. 156-160 °C; [α]_D -23 (chloroform); Ref.: 344

α-D-Galp-(1→6)-β-D-Galp-(1→6)-D-Gal

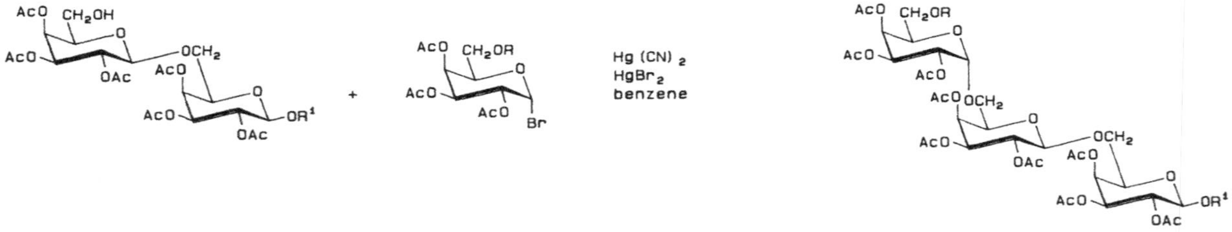

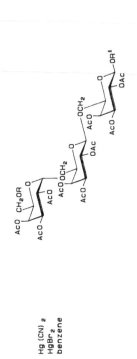

+

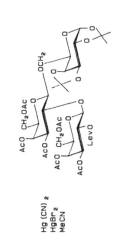

Hg(CN)₂
HgBr₂
benzene

R¹=Me; R=MCA; C-13; Ref.: 345

R¹=All; R=Ac; 10%; [α]ₒ +44 (chloroform); C-13; Ref.: 346

β-D-Galp-(1→2)-β-D-Galp-(1→6)-D-Gal

+

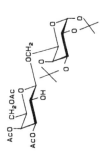

Hg(CN)₂
HgBr₂
MeCN

74%; Ref.: 344, 347

β-D-Galp-(1→6)-β-D-Galp-(1→6)-D-Gal

$[\alpha]_{589}$ +20 (water): Ref.: 348

β-D-(1→OMe): foam; $[\alpha]_D$ +38.1 (water): C-13: Ref.: 349
β-D-(1→OMe): m.p. 203-205°C; $[\alpha]_D$ -10 (water): C-13: Ref.: 345
β-D-(1-O-2,3-epoxypropyl): $[\alpha]_D$ -17.4 (water): C-13: Ref.: 346

I   TMSOTf

II AgOTf
s-Coll
dichloro-
methane
Ref.: 350

III Hg(CN)₂
HgBr₂
benzene
Ref.: 346

IV Ag₂CO₃
benzene
Ref.: 348

III Ref.: 345, 346

I   R¹=β-OMe; R=Bz; R²=MBA; 71%; m.p. 233-234°C;
$[\alpha]_D$ +162 (chloroform): C-13: Ref.: 350

II R¹=β-OMe; R=R²=Bz; 80%; m.p. 250-251°C;
$[\alpha]_D$ +115 (chloroform): C-13: Ref.: 350

III R¹=β-OAll; R=R²=Ac
70%; $[\alpha]_D$ -18.2 (chloroform): C-13: Ref.: 346

R¹=β-OAll; R²=MBA
75%; $[\alpha]_D$ -13.5 (chloroform): C-13: Ref.: 346

R¹=β-OMe; R=Ac; R²=MCA
58.7%; $[\alpha]_D$ -9 (chloroform): C-13: Ref.: 345

IV R¹=OAc; R=Ac; R²=MCA
25%; $[\alpha]_{589}$ +12 (chloroform); Ref.: 348

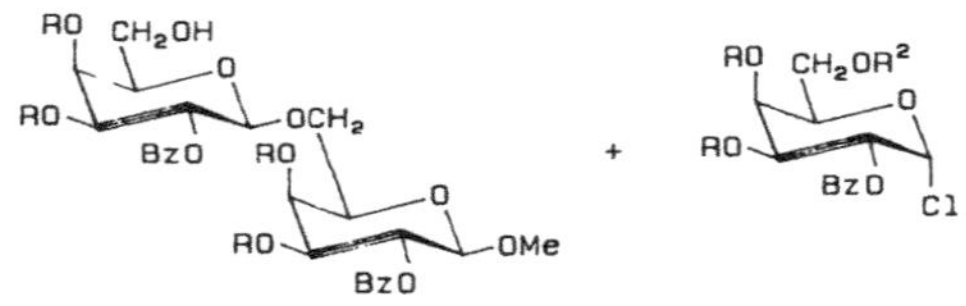

V   AgOTs
     MeCN

II   Ref.: 350, 351

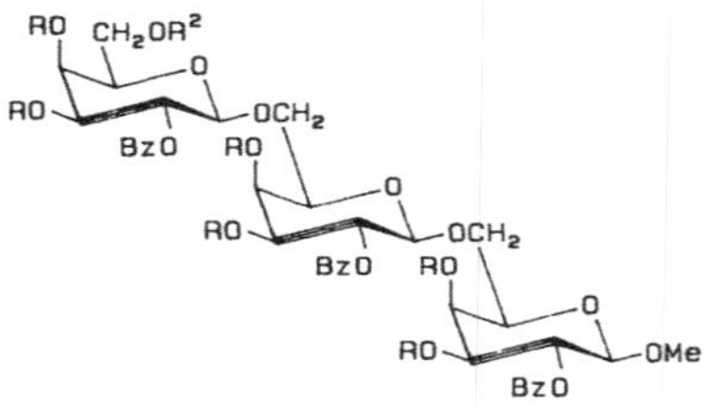

II   R=Bz; R²=MBA; 89%; Ref.: 350, 351

V   R=Bzl; R²=Ac; 84%; m.p. 84-86°C;
     [α]ᴅ +30.1 (chloroform); C-13: Ref.: 349

3-Deoxy-3-fluoro-α-D-Galp-(1→6)-α-D-Galp-(1→6)-D-Gal

α-D-(1-OMe): colourless foam; [α]ᴅ -14 (water); Ref.: 352
α-D-(1-OMe): m.p. 191-193°C; [α]ᴅ -14 (water); C-13: Ref.: 345

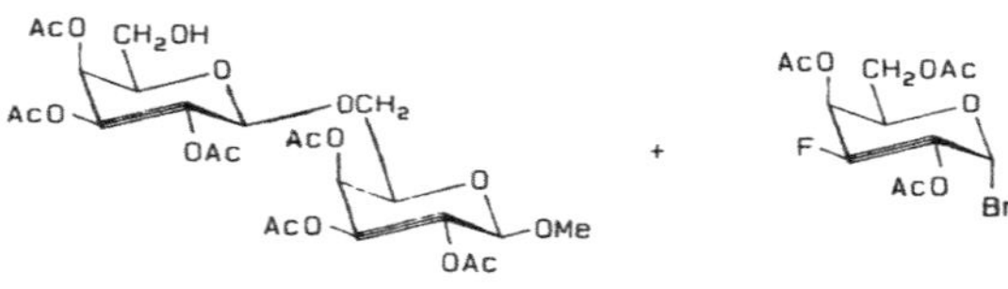

Hg(CN)₂
HgBr₂
benzene

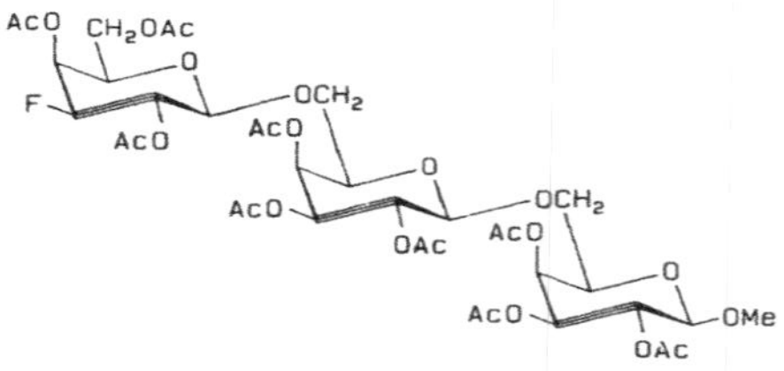

62-67%; [α]ᴅ -80.3 (chloroform); C-13: Ref.: 345, 352

β-D-Galp-(1→3)-α-D-GalpNAc-(1→6)-D-Gal

[α]$_D$ -10.1 (water): Ref.: 124

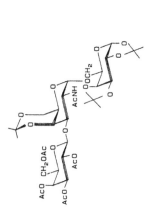

Hg(CN)$_2$
HgBr$_2$
toluene
MeNO$_2$

71.6%; m.p. 102°C; [α]$_D$ +5.3 (chloroform); Ref.: 124

α-D-GalpN-(1→3)-β-D-GalpNAc-(1→3)-D-Gal
α-D-GalpNAc-(1→3)-β-D-GalpNAc-(1→3)-D-Gal

$[\alpha]_D$ +53.2 (water): Ref.: 353

I   Ag$_2$CO$_3$ dichloro-methane

II   AgOTf s-Coll

III   Ag-silicate dichloro-methane

IV   Ag$_2$CO$_3$ AgClO$_4$ dichloro-methane

V   AgClO$_4$ polyvinyl-pyridine dichloro-methane

I   R^1=N$_3$; R=Bzl; X=α-Br
33%; $[\alpha]_D$ +27.4 (chloroform): Ref.: 354

II   R^1=NPhth; R=Bz; X=Br
68%; $[\alpha]_D$ +87.6: Ref.: 355

III   R^1=NPhth; R=Bz; X=Br
67.5%; $[\alpha]_D$ +87.6 (chloroform): Ref.: 104

IV   R^1=N$_3$; R=Bzl
58%; $[\alpha]_D$ +31.9 (dichloromethane): Ref.: 353

     R^1=NHAc; R=Bz
63%; $[\alpha]_D$ +58.9 (dichloromethane): Ref.: 353

V   R^1=N$_3$; R=Bz
28%; $[\alpha]_D$ +82.0 (dichloromethane): Ref.: 353

β-D-GalpN- (1→3) -β-D-GalpN- (1→3) -D-Gal

Ag₂CO₃
AgClO₄
dichloro-
methane

65%; [α]_D +23.2 (chloroform); Ref.: 353

β-D-Xylp- (1→3) 〉D-Gal
β-D-Xylp- (1→4)

Hg(CN)₂
benzene
MeNO₂

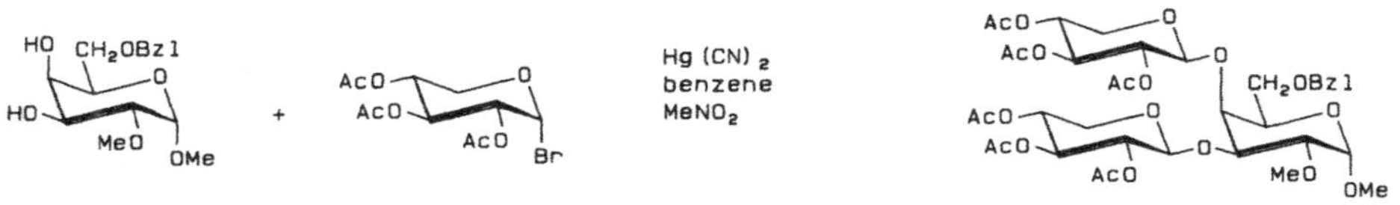

45%; [α]_D -0.3 (chloroform); Ref.: 356

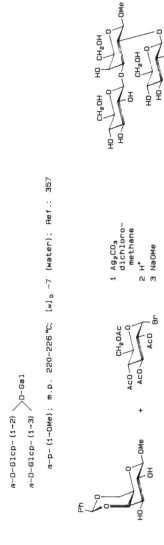

α–L–Rhap–(1→3) ⟩D–Gal
α–D–Glcp–(1→6)

[α]$_D$ +50.8 (water); C–13; Ref.: 323

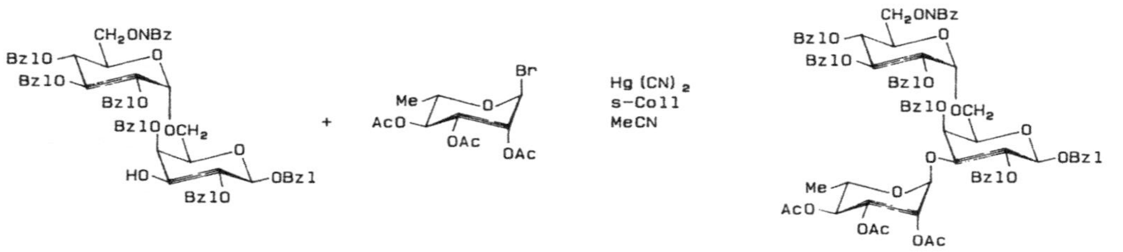

~100%; [α]$_D$ +11 (chloroform); Ref.: 323

Hg(CN)$_2$
s–Coll
MeCN

β–D–Glcp–(1→2) ⟩D–Gal
α–D–Glcp–(1→3)

β–D–(1→OMe); m.p. 220–226°C; [α]$_D$ –7 (water); Ref.: 357

1 Ag$_2$CO$_3$
  dichloro-
  methane
2 H$^+$
3 NaOMe

4.7%; Ref.: 357

β-D-Glcp-(1→2)
β-D-Galp-(1→6) >D-Gal

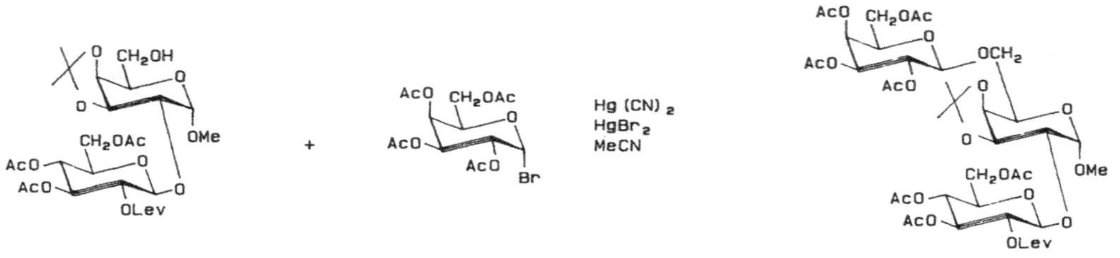

+

Hg(CN)₂
HgBr₂
MeCN

Ref.: 358

β-D-Glcp-(1→3)
β-D-Glcp-(1→4) >D-Gal

m.p. 126—130 °C;   [α]_D +27.5 (water); Ref.: 359

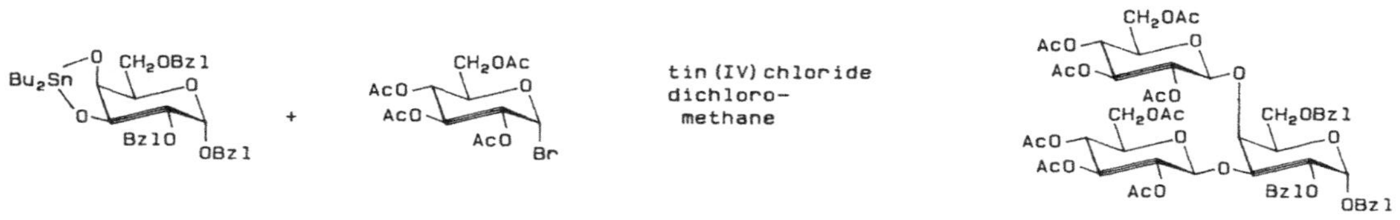

+

tin(IV)chloride
dichloro-
methane

25%; [α]_D +23 (chloroform); Ref.: 359

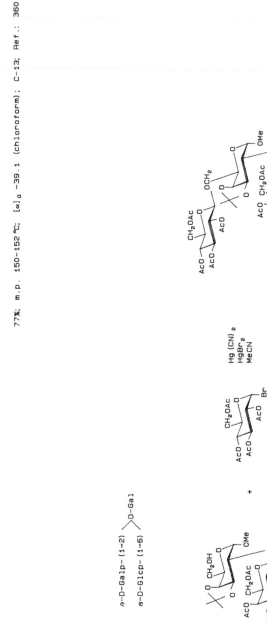

α-L-Rhap-(1→2)
                ⟩D-Gal
β-D-Glcp-(1→3)

m.p. 195—197 °C; [α]$_D$ -7.2 → -4.5 (water); Ref.: 360

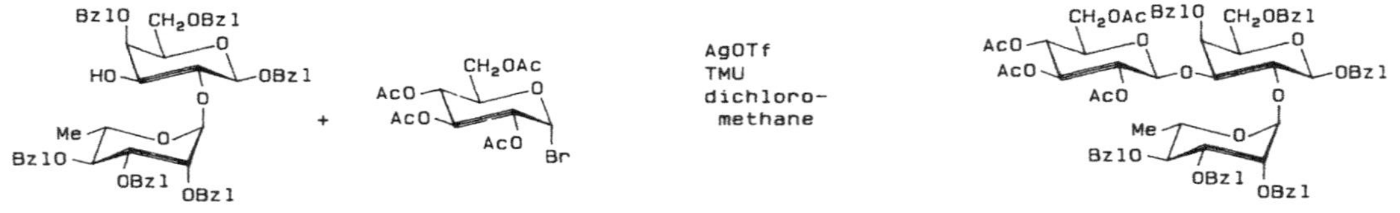

AgOTf
TMU
dichloro-
methane

77%; m.p. 150—152 °C; [α]$_D$ -39.1 (chloroform); C-13; Ref.: 360

β-D-Galp-(1→2)
                ⟩D-Gal
β-D-Glcp-(1→6)

Hg(CN)$_2$
HgBr$_2$
MeCN

70%; m.p. 201 °C; [α]$_D$ +36.2 (chloroform); C-13; Ref.: 344

α-D-GlcpN-(1→3)
                 \ D-Gal
β-D-Manp-(1→4) /

α-D-GlcpNAc-(1→3)
                   \ D-Gal
β-D-Manp-(1→4) /

m.p. 181 °C; [α]_D +92.7 (water); Ref.: 298

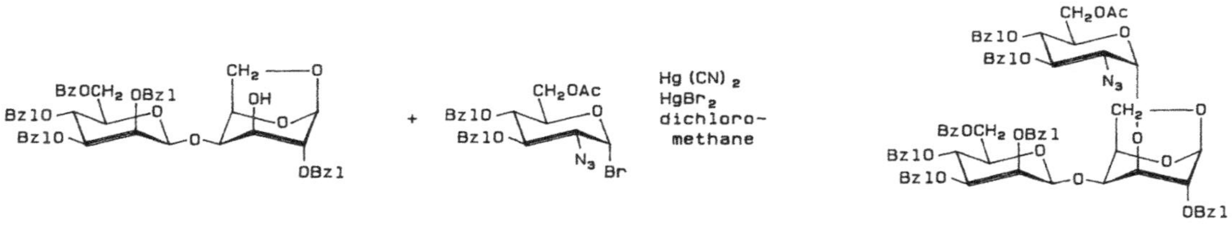

76%; [α]_D −1.7 (dichloromethane); Ref: 298

Hg(CN)₂
HgBr₂
dichloro-
methane

β-D-GlcpNAc-(1→3)
                    \ D-Gal
β-D-GlcpNAc-(1→4) /

m.p. 170—174 °C; [α]_D +24 (water); Ref.: 361

pTSA
MeNO₂
toluene

9%; m.p. 182—183 °C; [α]_D +10 (chloroform); Ref.: 361

β-D-GlcpNAc-(1→3)
⟩D-Gal
β-D-GlcpNAc-(1→6)

m.p. 142—144°C (dec.); [α]$_D$ +6.5 (water); Ref.: 362

50%; m.p. 228°C (dec.); [α]$_D$ −35.2 (chloroform); Ref.: 362

β-D-GlcpN-(1→3)
⟩D-Gal
β-D-Manp-(1→4)

β-D-GlcpNAc-(1→3)
⟩D-Gal
β-D-Manp-(1→4)

[α]$_D$ −3.8 (MeOH); Ref.: 363

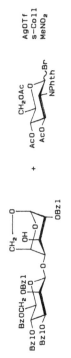

AgOTf
s-Coll
MeNO$_2$

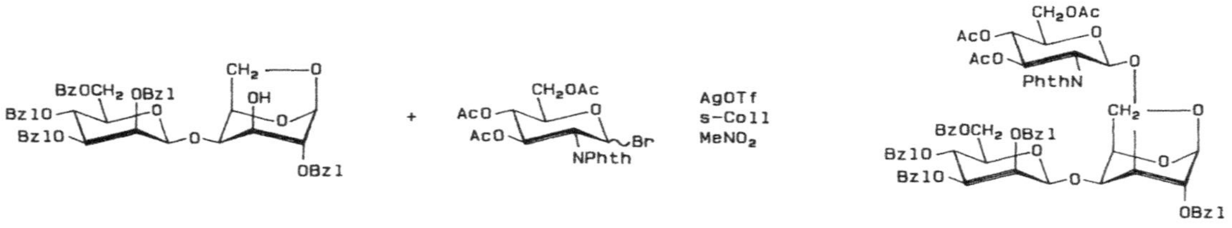

92.7%; [α]$_D$ −41.6 (chloroform); Ref: 363

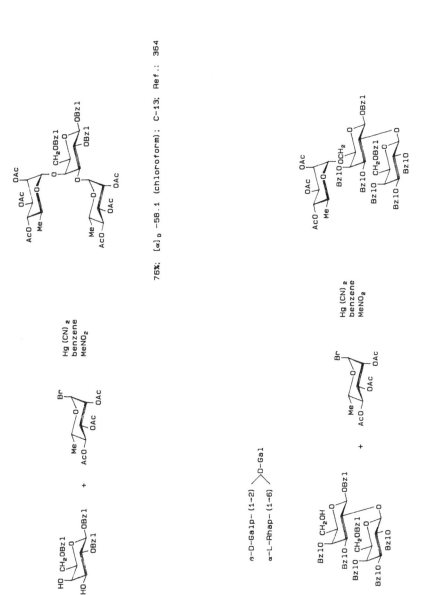

α-L-Rhap-(1→3)
                    ⟩D-Gal
α-L-Rhap-(1→4)

foam: [α]$_D$ −14.4 (water): C-13: Ref.: 364

76%: [α]$_D$ −58.1 (chloroform): C-13: Ref.: 364

α-D-Galp-(1→2)
                    ⟩D-Gal
α-L-Rhap-(1→6)

76%: m.p. 52 °C: [α]$_D$ −34.1 (chloroform): Ref.: 365

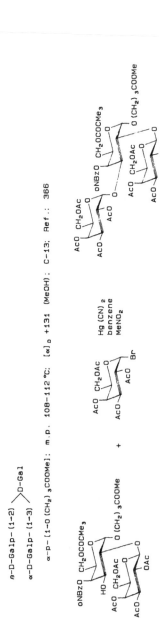

β-D-Galp-(1→2)
                    ⟩D-Gal
α-D-Galp-(1→3)

α-p-[1-O(CH₂)₃COOMe]: m.p. 108-112 °C; [α]$_D$ +131 (MeOH); C-13; Ref.: 366

34%; m.p. 76-79 °C; [α]$_D$ +20.7 (chloroform); C-13; Ref.: 366

α-L-Fucp-(1→2)
                \
α-D-Galp-(1→3) / D-Gal

[α]$_D$ +35.2 (water): Ref.: 367

foam; [α]$_D$ +38 (water): Ref.: 359

m.p. 135-138 °C; [α]$_D$ +35 (water-MeOH. 19:1): Ref.: 368

[α]$_D$ +35 (water): Ref.: 369

[α]$_D$ -34 (water): Ref.: 370

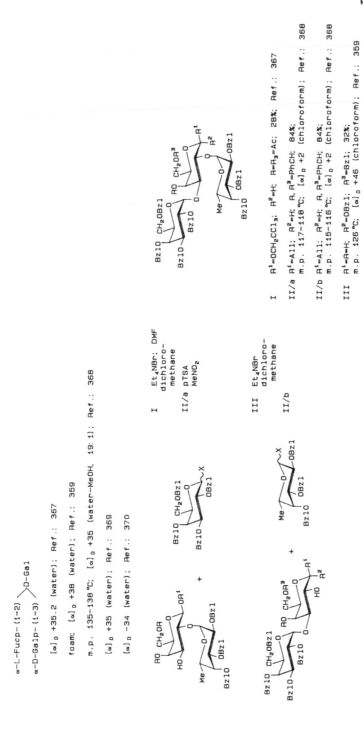

I   Et$_4$NBr: DMF
      dichloro-
      methane

II/a pTSA
      MeNO$_2$

III   Et$_4$NBr
      dichloro-
      methane

II/b

I   R^1=OCH$_2$CCl$_3$; R^2=H; R=R$_3$=Ac: 28%; Ref.: 367

II/a R^1=All; R^2=H; R, R^3=PhCH; 84%;
      m.p. 117-118 °C; [α]$_D$ +2 (chloroform); Ref.: 368

II/b R^1=All; R^2=H; R, R^3=PhCH; 84%;
      m.p. 115-116 °C; [α]$_D$ +2 (chloroform); Ref.: 368

III   R^1=R=H; R^2=OBzl; R^3=Bzl; 32%;
      m.p. 126 °C; [α]$_D$ +46 (chloroform); Ref.: 359

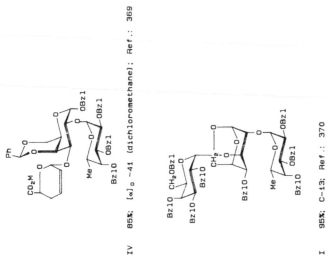

IV   85%; [α]$_D$ −41 (dichloromethane); Ref.: 369

I   95%; C-13; Ref.: 370

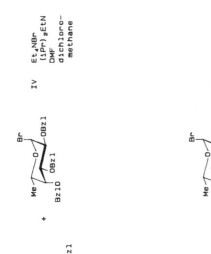

α-L-Fucp-(1-2)
2-OAc-α-D-Galp-(1-3) ⟩ D-Gal

[α]_D +66 (water): Ref.: 369

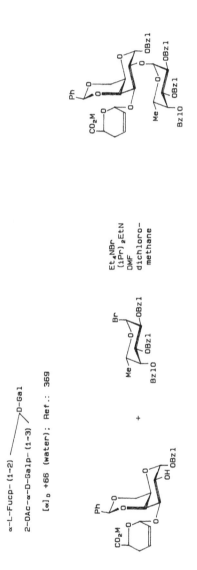

+

Et₄NBr
(iPr)₂EtN
DMF
dichloro-
methane

85%: [α]_D -41 (dichloromethane): Ref.: 369

α-D-Galp-(1-3) ⟩ D-Gal
α-L-Fucp-(1-4)

foam; [α]_D +45 (water): Ref.: 359

+

Et₄NBr
dichloro-
methane

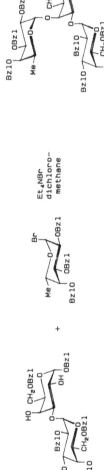

11%: [α]_D +2.5 (chloroform): Ref.: 359

β-D-Galp-(1→2)
β-D-Galp-(1→3) ⟩D-Gal

α-p-[1→O(CH₂)₃COOMe]: m.p. 128-130 °C; [α]ᴅ +57.9 (MeOH); C-13; Ref.: 366

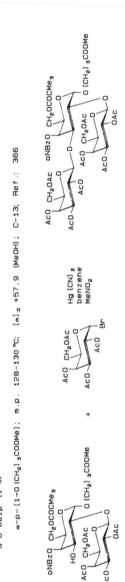

34%; m.p. 80-83 °C; [α]ᴅ -5.3 (chloroform); C-13; Ref.: 366

α-L-Fucp-(1→2)
α-D-GalN-(1→3) ⟩D-Gal

decomp.; Ref.: 369

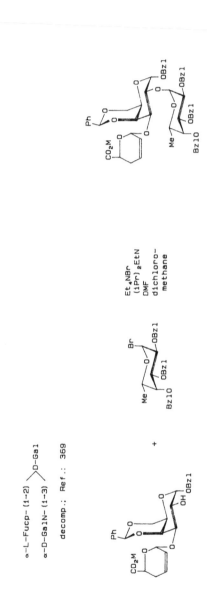

85%; [α]ᴅ -41 (dichloromethane); Ref.: 369

α—L—Fucp—(1→2) ⟩ D—Gal

α—D—GalpNAc—(1→3)

m.p. 143—148 °C; [α]_D +36.5 (water): Ref.: 369, 371

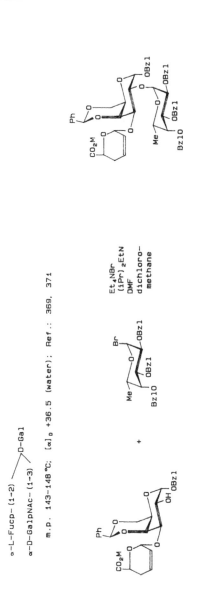

+

Et₄NBr
(iPr)₂EtN
DMF
dichloro—
methane

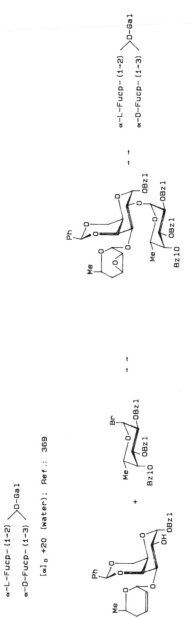

85%; [α]_D −41 (dichloromethane): Ref.: 369, 371

α—L—Fucp—(1→2) ⟩ D—Gal

α—D—Fucp—(1→3)

[α]_D +20 (water): Ref.: 369

α—L—Fucp—(1→2) ⟩ D—Gal

α—D—Fucp—(1→3)

[α]_D +26 (dichloromethane): Ref.: 369

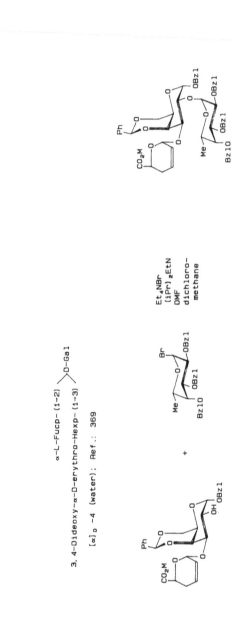

α-L-Fucp-(1-2)
                    ⟩D-Gal
3-Deoxy-α-D-xylo-Hexp-(1-3)

[α]_D +24 (water): Ref.: 369

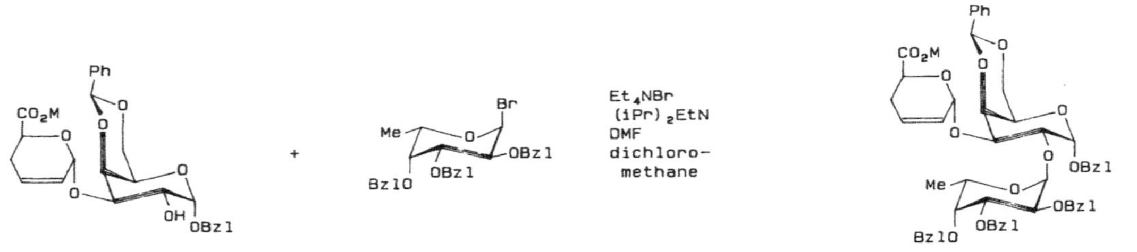

+

Et₄NBr
(iPr)₂EtN
DMF
dichloro-
methane

85%: [α]_D -41 (dichloromethane): Ref.: 369

α-L-Fucp-(1-2)
                    ⟩D-Gal
3.4-Dideoxy-α-D-erythro-Hexp-(1-3)

[α]_D -4 (water): Ref.: 369

+

Et₄NBr
(iPr)₂EtN
DMF
dichloro-
methane

85%: [α]_D -41 (dichloromethane): Ref.: 369

β-D-Galp-(1→4)-β-D-Glcp-(1→3)-D-GalN

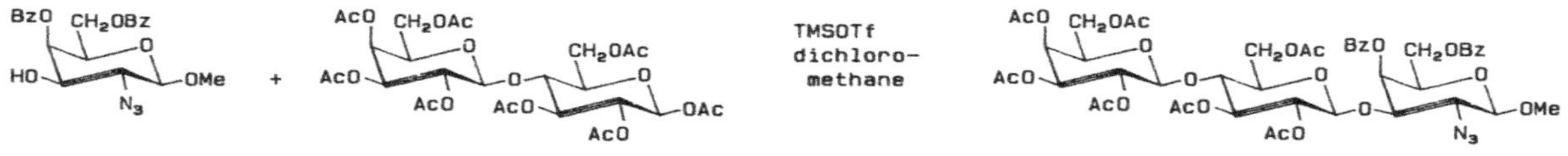

TMSOTf
dichloro-
methane

66%; [α]_D -25 (chloroform): Ref.: 372

β-D-Galp-(1→4)-β-D-GlcpN-(1→3)-D-GalN

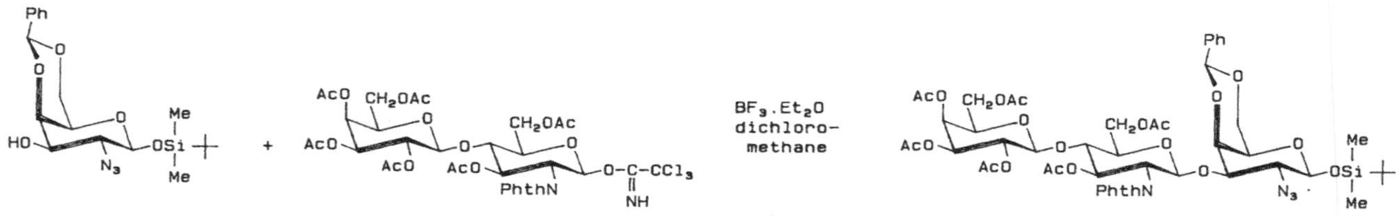

BF₃·Et₂O
dichloro-
methane

67%; [α]_578 +18 (chloroform): C-13: Ref.: 373

β-D-Galp-(1→4)-β-D-GlcpN-(1→6)-D-GalN

β-D-Galp-(1→4)-β-D-GlcpNAc-(1→6)-D-GalNAc

α-D-(1-OMCO): powder; [α]_D +129 (water); C-13; Ref.: 374

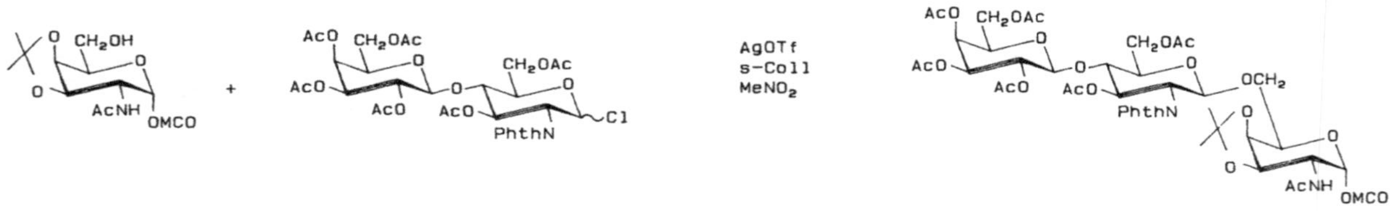

75%; m.p. 102—105 °C; [α]_D +57.3 (chloroform); C-13; Ref.: 374

α-L-Fucp-(1→2)-α-D-Galp-(1→3)-D-GalNAc

α-p-(1→OPh): [α]$_D$ +49.4 (MeOH): Ref.: 375

α-p-(1-O-o-nitrophenyl): m.p. 290-291 °C; [α]$_D$ +161.9 (DMSO): Ref.: 180

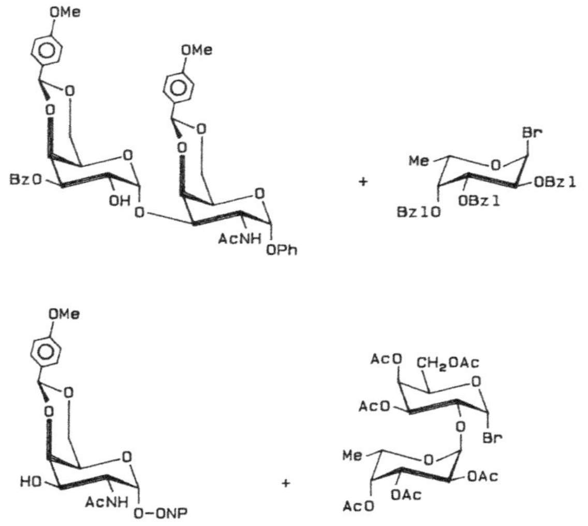

I   Et$_4$NBr
    DMF
    dichloro-
    methane

II 1 Hg(CN)$_2$
     benzene
     MeNO$_2$
   2 AcOH

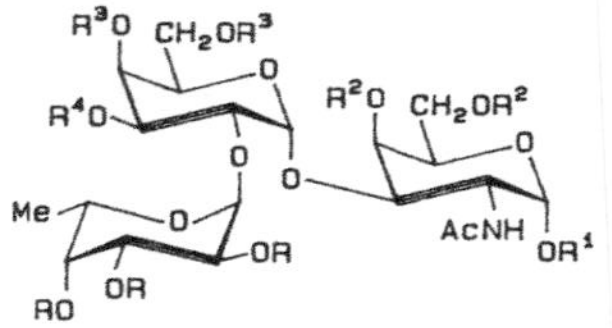

I   R^1=Ph; R^2, R^2=p-MeO-PhCH;
    R^3, R^3=p-MeO-PhCH; R^4=Bz; R=Bzl
    79%; m.p. 205-206 °C;
    [α]$_D$ +49 (chloroform); Ref.: 375

II R^1=ONP; R^2=H; R^3=R^4=R=Ac
    51.4%; [α]$_D$ +95.9 (MeOH); C-13; Ref.: 180

β-D-GlcpNAc-(1→3)-α-D-Galp-(1→3)-D-GalNAc

α-p-(1-OBzl): m.p. 300 °C (dec.): [α]$_D$ +68.6 (DMF): Ref.: 376

+

Hg(CN)$_2$
benzene
MeNO$_2$

41.2%: [α]$_D$ +86.2 (chloroform): Ref.: 376

β-D-Galp-(1→3)-β-D-Galp-(1→3)-D-GalN
β-D-Galp-(1→3)-β-D-Galp-(1→3)-D-GalNAc

α-p-(1-OMCO): amorphous powder: [α]$_D$ +23.2 (MeOH): Ref.: 377

TMSOTf
dichloro-
methane

45%: [α]$_D$ +46.6 (chloroform): Ref.: 377

β-D-GalpN-(1→3)-β-D-Galp-(1→3)-D-GalN

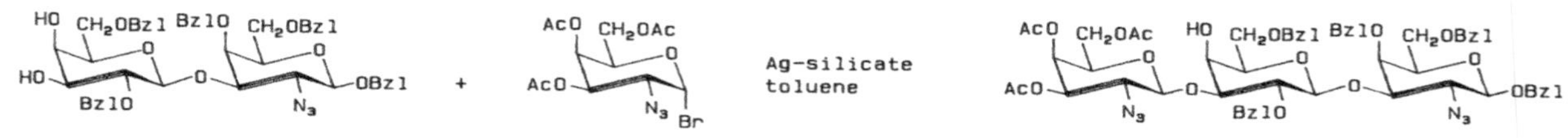

11%; [α]$_D$ -21 (chloroform): Ref.: 378

Ag-silicate
toluene

+

[α]$_D$ +14 (water): Ref.: 378

β-D-GalpN-(1→4)-β-D-Galp-(1→3)-D-GalN

β-D-GalpNAc-(1→4)-β-D-Galp-(1→3)-D-GalNAc

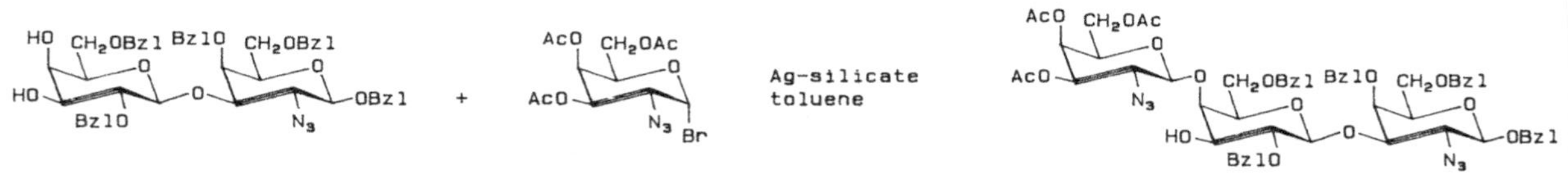

60%; [α]$_D$ -16.1 (chloroform): Ref.: 379

61%; [α]$_D$ -16.1 (chloroform); C-13: Ref.: 378

Ag-silicate
toluene

+

α-D-GalpN-(1→3)-β-D-Galp-(1→4)-D-GalN

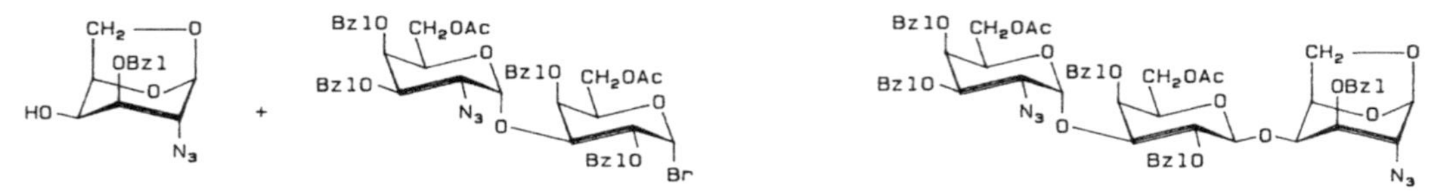

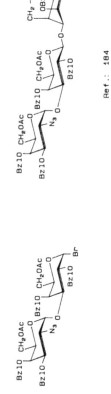

Ref.: 184

α-L-Fucp-(1→2)-β-D-Galp-(1→4)-D-GalNAc

β-p-(1→OPr): [α]$_D$ -66.4 (MeOH): C-13; Ref.: 380

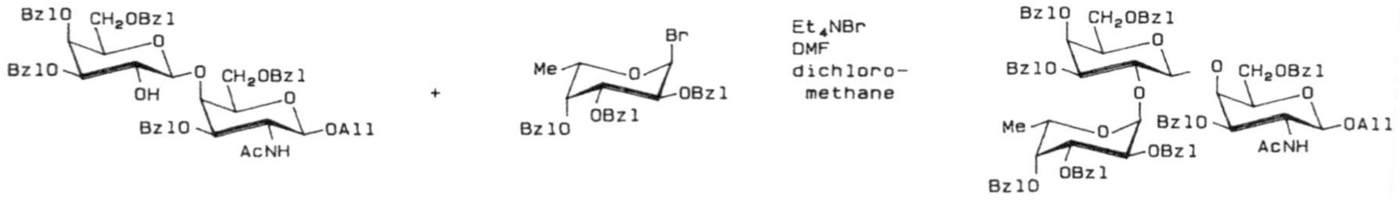

Et$_4$NBr
DMF
dichloro-
methane

74%; [α]$_D$ -34.3 (chloroform); Ref.: 380

β-D-ManpN-(1→3)-α-L-FucpN-(1→3)-D-GalN

β-D-ManpNAc-(1→3)-α-L-FucpNAc-(1→3)-D-GalNAc

syrup: [α]_D -93.0 → -89.4 (MeOH); C-13; Ref.: 381

 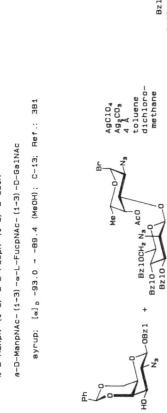

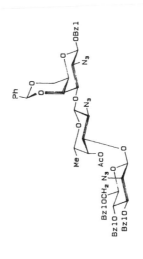

64%; [α]_D -74.6 (chloroform); Ref.: 381

β-D-ManpN-(1→3)-α-L-FucpN-(1→3)-D-GalN

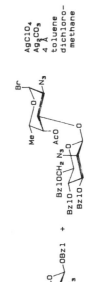

14%; [α]_D -35.5 (chloroform); Ref.: 381

α—D-GalpN-(1-3)
    ⟩D-GalN
α—D-GlcpN-(1-4)

α—D-GalpNAc-(1-3)
    ⟩D-GalNAc
α—D-GlcpNAc-(1-4)

amorphous; [α]$_D$ +123 (water); Ref.: 382

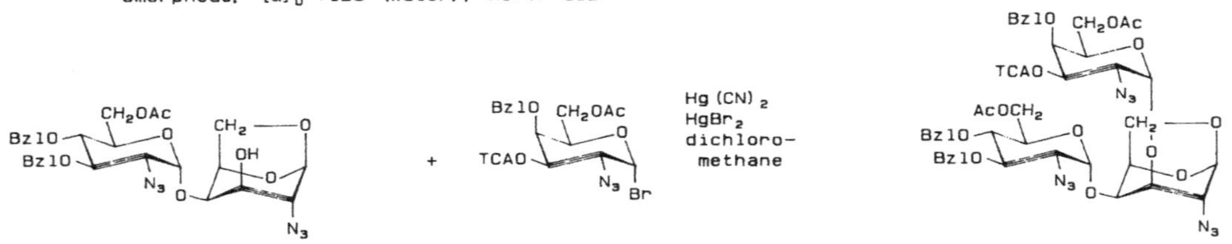

64%; [α]$_D$ +116 (dichloromethane); Ref: 382, 383

β-D-GlcpN-(1→3)
                 ⟩D-GalN
β-D-GlcpN-(1→6)

β-D-GlcpNAc-(1→3)
                   ⟩D-GalNAc
β-D-GlcpNAc-(1→6)

β-p-(1-OBzl): amorphous; $[\alpha]_D$ +66.3 (water); C-13; Ref.: 384

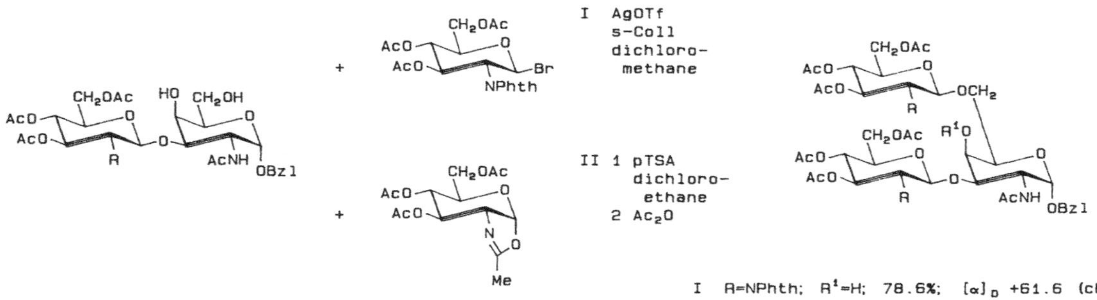

I   AgOTf
    s-Coll
    dichloro-
    methane

II 1 pTSA
    dichloro-
    ethane
    2 Ac₂O

I   R=NPhth; R¹=H; 78.6%; $[\alpha]_D$ +61.6 (chloroform); Ref.: 384

II R=NHAc; R¹=Ac; 62.5%; $[\alpha]_D$ +51.7 (chloroform); Ref.: 384

β-D-Galp-(1→3)
β-D-GlcpNAc-(1→6) ⟩ →D-GalNAc

α-p-(1-OBzl): m.p. 254—257 °C; [α]$_D$ +70 (water); C-13; Ref.: 385

AcO CH$_2$OAc HO CH$_2$OH
AcO

OAC AcNH OBzl

+

CH$_2$OAc
AcO
AcO
N
O
Me

pTSA
dichloro-
ethane

CH$_2$OAc
AcO
AcO
OCH$_2$
AcNH
AcO CH$_2$OAc
AcO
OAC
ACNH OBzl
HO

75.9%: [α]$_D$ +31.2 (chloroform); Ref.: 385

β-D-Fucp-(1→3)
β-D-GlcpNAc-(1→6) ⟩ →D-GalNAc

α-p-(1-OBzl): m.p. 282—284 °C (dec.); [α]$_D$ +66.4 (water); C-13; Ref.: 386

AcO Me CH$_2$OH
AcO
HO

OAC AcNH OBzl

+

CH$_2$OAc
AcO
AcO
N
O
Me

pTSA
dichloro-
ethane

CH$_2$OAc
AcO
AcO
OCH$_2$
AcNH
AcO Me
AcO
OAC
ACNH OBzl
HO

47%: m.p. 268—270 °C; [α]$_D$ +45.4 (MeOH); Ref.: 386

β-D-Galp-(1→3)
$\qquad\qquad\qquad$ \\ D-GalN
β-D-Galp-(1→4) /

amorphous: [α]_D +34.3 (water); C-13; Ref.: 388, 389

α-(1-O-Sp); amorphous glass; [α]_D +61 (water); Ref.: 387
Sp=-(CH2)2NH-CO-(CH2)4COOMe

I   TMSOTf

II 1 TMSOTf dichloro-
    methane
    2 NaOMe
    3 Ac2O

III AgOTf
    TMU
    dichloro-
    methane

IV Hg(CN)2
    HgBr2
    benzene
    MeNO2

III

V   Hg(CN)2
    HgBr2
    MeNO2

III

I   R¹=H; R²=OMe; R³=N3; R=Bz; Ref.: 390

II   R¹=H; R²=OMe; R³=N3; R=Ac; 51%
     [α]_D -32 (chloroform); Ref.: 372

III R¹=OBzl; R²=H; R³=NHAc; R=Bzl; 21%; Ref.: 387

     [2+1] R¹=OBzl; R²=H; R³=NHAc; R=Bzl; 69%
     m.p. 194-195°C; [α]_D +33.6 (chloroform); Ref.: 387

IV   R¹=O-Sp; R²=H; R³=NHAc; R=Bz; 76%
     m.p. 149°C; [α]_D +22 (chloroform); Ref.: 388, 389

V   R¹=O-Sp; R²=H; R³=NHAc; R=Ac; 40%
     [α]_D +31 (chloroform); Ref.: 388, 389

β-D-Glcp-(1-2)
β-D-Glcp-(1-3) ⟩ L-Fuc

α-p-(1-OMe): m.p. 143-145 °C; [α]_D -57.1 (MeOH); C-13; Ref.: 305, 391

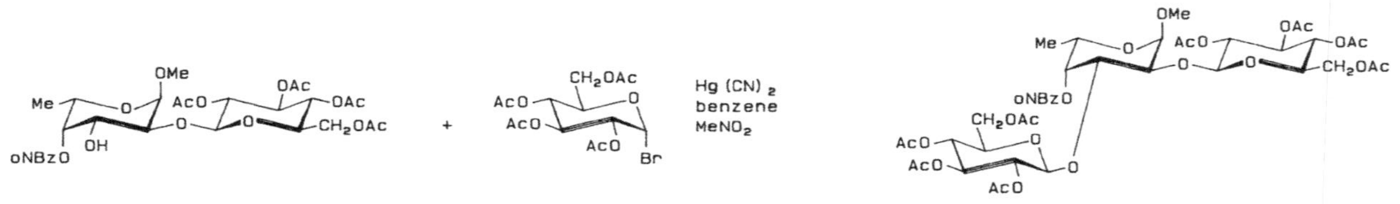

Hg(CN)_2
benzene
MeNO_2

60-75%; m.p. 92-94 °C; [α]_D -12 (chloroform); C-13; Ref.: 305, 391

β-D-Galp-(1-2)
β-D-Galp-(1-3) ⟩ L-Fuc

α-p-(1-OMe): m.p. 157-159 °C; [α]_D -58.2 (MeOH); C-13; Ref.: 305
α-p-(1-OMe): m.p. 145-147 °C; [α]_D -26.6 (MeOH); C-13; Ref.: 391

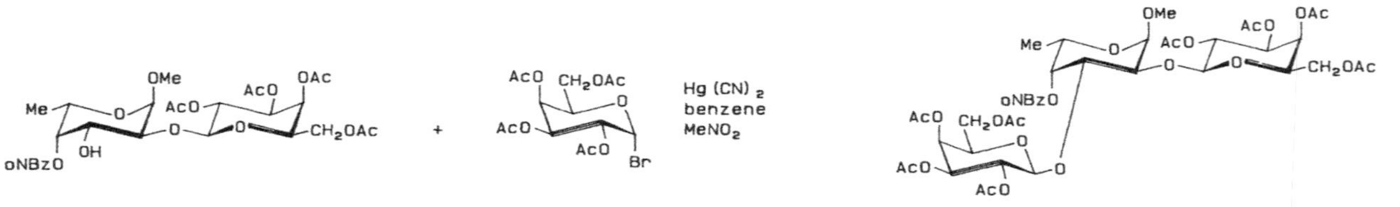

Hg(CN)_2
benzene
MeNO_2

80%; m.p. 95-96 °C; [α]_D -8.5 (chloroform); C-13; Ref.: 391

β-D-Glcp-(1→6)-α-D-Glcp-(1→2)-β-D-Fruf

m.p. 212 °C; [α]_D +30.8 (water); Ref.: 394

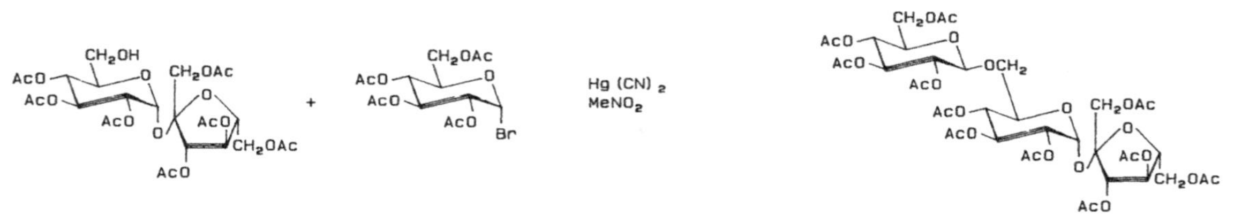

Hg(CN)_2
MeNO_2

69%; m.p. 80—93 °C; [α]_D +32.2 (chloroform); Ref: 394

α-D-Galp-(1→6)-α-D-Glcp-(1→2)-β-D-Fruf

m.p. 80—82 °C; [α]_D +105 (water); Ref.: 395

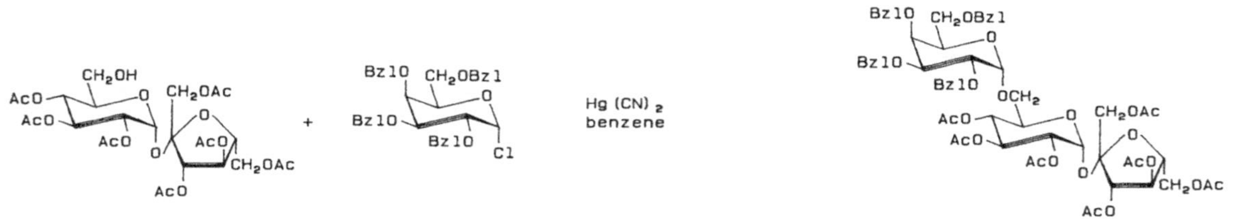

Hg(CN)_2
benzene

53%; [α]_D +65.4 (chloroform); Ref: 395

β-D-Galp-(1→6)-α-D-Glcp-(1→2)-β-D-Fruf

m.p. 132—135 °C; $[\alpha]_D$ +55.6 (water); Ref.: 395

I   Hg (CN)$_2$
    benzene

II Hg (CN)$_2$
    MeNO$_2$

I   R=Bzl; X=Cl; 8%; $[\alpha]_D$ +47.8 (chloroform); Ref: 395

II R=Ac; X=Br; 59%; m.p. 75—77 °C; $[\alpha]_D$ +46.5 (chloroform); Ref.: 395

β-D-Ribf-(1→2)-β-D-Ribf-(1→7)-KDO

α-(2-OMe) Na-salt; colourless glass; [α]$_D$ -10.4 (water); Ref.: 397

β-(2-OMe) Na-salt; colourless glass; [α]$_D$ -16.5 (water); Ref.: 397

AgOTf
dichloro-
methane

R¹=OMe; R=COOMe; 43%; [α]$_D$ +7.1 (chloroform); Ref.: 397
R¹=COOMe; R=OMe; 22%; [α]$_D$ +13.4 (chloroform); Ref.: 397

R¹=OMe; R=COOMe; 397
R¹=COOMe; R=OMe; 397

β-D-Galp-(1→6)-β-D-Galp-(1→6)-3-deoxy-3-flouro-D-Gal

β-p-(1→OMe); m.p. 215–217 °C; $[\alpha]_D$ −12.3 (water); C-13; Ref.: 400

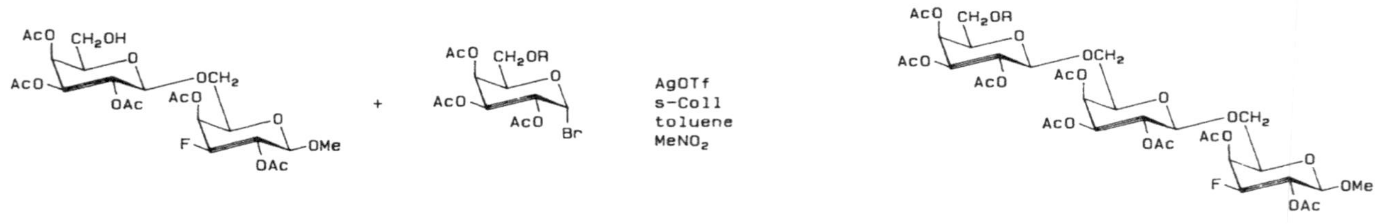

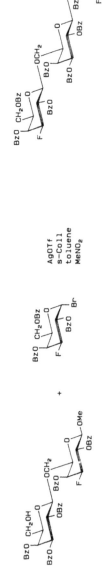

AcO
CH₂OH
OCH₂
OAc
F
OMe
AcO
CH₂OR
Br
AgOTf
s-Coll
toluene
MeNO₂

R=Ac; 42%; $[\alpha]_D$ −9.1 (chloroform); C-13; Ref.: 400

R=MBA; 41.3%; m.p. 128–130 °C; $[\alpha]_D$ −6.6 (chloroform); C-13; Ref.: 400

3-Deoxy-3-fluoro-β-D-Galp-(1→6)-β-D-Galp-(1→6)-3-deoxy-3-flouro-D-Gal

β-p-(1→OMe); m.p. 233–235 °C; $[\alpha]_D$ +14.2 (water); C-13; Ref.: 401

BzO
CH₂OH
OCH₂
OBz
F
OMe
CH₂OBz
Br
AgOTf
s-Coll
toluene
MeNO₂

93%; $[\alpha]_D$ +104.6 (chloroform); C-13; Ref.: 401

β-D-ManpNA-(1→4)-α-D-GlcpN-(1→3)-D-Fucp4N

β-D-ManpNAcA-(1→4)-α-D-GlcpNAc-(1→3)-D-Fucp4NAc

α-p-(1-OMe): amorphous: $[\alpha]_D$ +74.0 (water): +84.6 (MeOH): C-13: Ref.: 392

α-p-(1-OMCO)-Na-salt: amorphous: $[\alpha]_D$ +74.3 (water): +83.6 (MeOH): C-13: Ref.: 392, 393

Ag$_2$CO$_3$
AgClO$_4$
dichloro-
methane
X=Cl: Ref.: 392, 393
X=Br: Ref.: 392

R_1=MCO: R=Bzl: 40%: $[\alpha]_D$ +7.6 (chloroform): Ref.: 392, 393

R_1=R=Me: $[\alpha]_D$ +4.4 (chloroform)
34% (X=Cl): Ref.: 392
40% (X=Br): Ref.: 392

β-D-ManpNA-(1→4)-β-D-GlcpN-(1→3)-D-Fucp4N

Ag$_2$CO$_3$
AgClO$_4$
dichloro-
methane
X=Cl: Ref.: 392, 393
X=Br: Ref.: 392

R_1=MCO: R=Bzl: 13%: Ref.: 392, 393

R_1=R=Me:
14% (X=Cl): Ref.: 392
14% (X=Br): Ref.: 392

β-D-Glcp-(1→4)-2,6-dideoxy-2-I-α-D-Altp-(1→4)-6-deoxy-D-All

NIS
MeCN

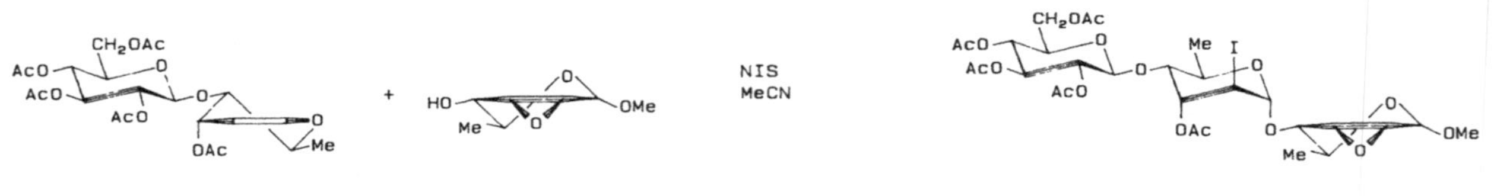

29%; syrup; [α]_D +58 (chloroform); Ref.: 398

2,6-Dideoxy-α-D-ribo-Hexp-(1→4)-2,6-dideoxy-2-I-α-D-Altp-(1→4)-6-deoxy-D-All

NIS
MeCN

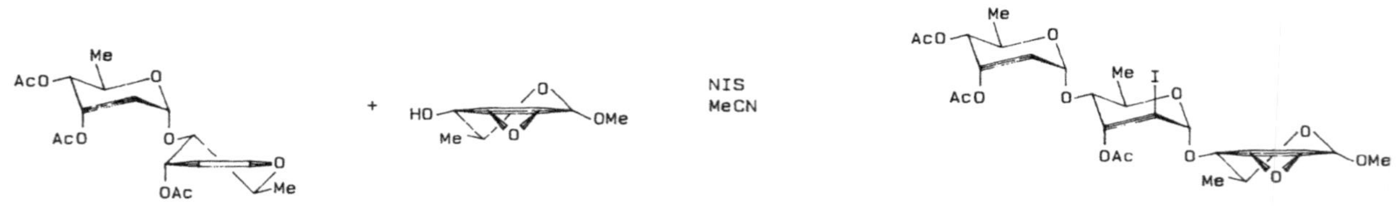

53%; syrup; [α]_D +233.3 (dichloromethane); Ref.: 399

2,6-Dideoxy-ß-D-ribo-Hexp-(1→4)-2,6-dideoxy-ß-D-ribo-Hexp-(1→4)-2,6-dideoxy-D-ribo-Hex

α-(1→OR): m.p. 214—216 °C; [α]$_D$ +8.65 (chloroform): Ref.: 402

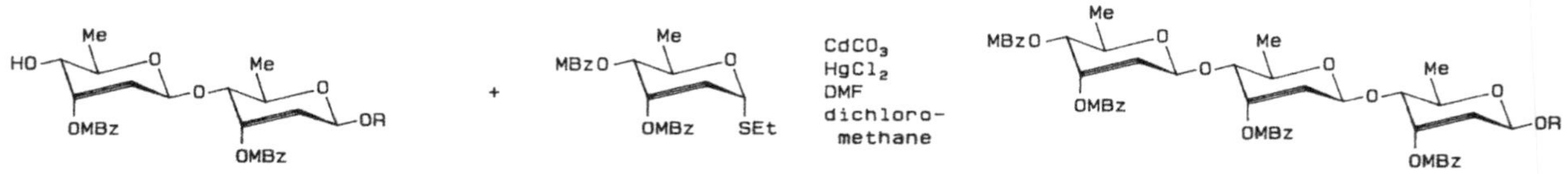

CdCO$_3$
HgCl$_2$
DMF
dichloro-
methane

+

57.8%: m.p. 159—161 °C; [α]$_D$ +112.9 (chloroform): Ref.: 402

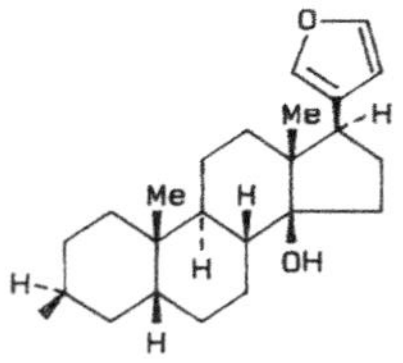

MBz=p-Methoxybenzoyl

2-Deoxy-2-I-α-D-Manp-(1-3) ⟩ 2,6-dideoxy-D-ribo-Hexp
2-Deoxy-2-I-α-D-Manp-(1-4) ⟩

NIS
MeCN

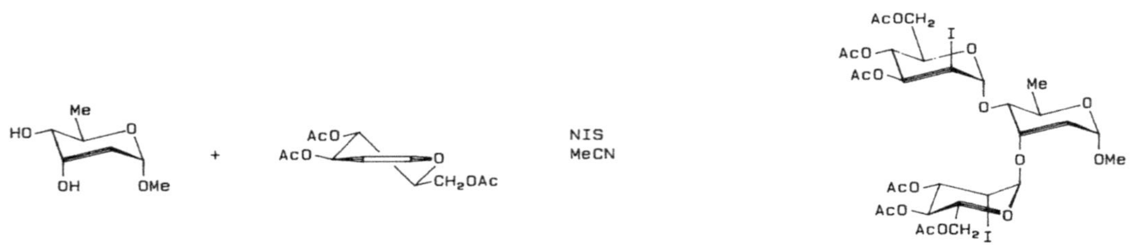

B%; syrup; [α]_D +74.3 (chloroform); Ref.: 403

2,6-Dideoxy-3-C-methyl-α-L-arabino-Hexp-(1→3)-2,6-dideoxy-β-D-arabino-Hexp-(1→3)-2,6-dideoxy-D-arabino-Hex

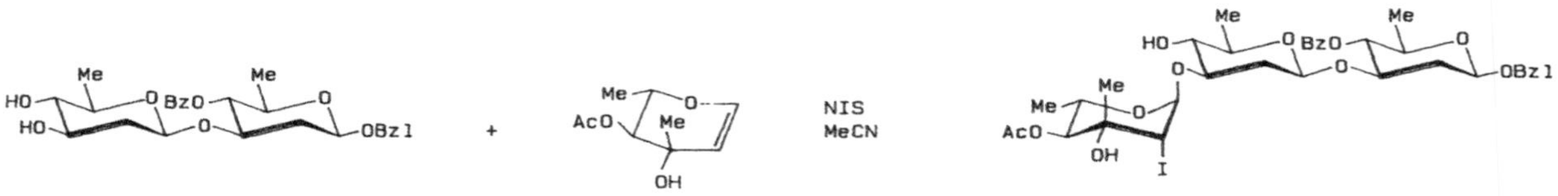

62%; [α]_D -44.5 (EtOAc); Ref.: 404

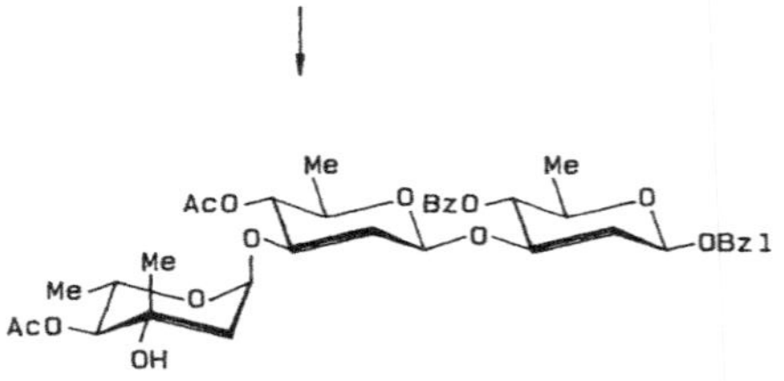

[α]_D -16.6 (EtOAc); Ref.: 404

2-Deoxy-α-L-Fucp-(1→4)-2-deoxy-α-L-Fucp-(1→4)-L-Rhodosamine

70%; m.p. 64°C; $[\alpha]_D$ -145 (chloroform); Ref.: 405

α-L-Cinerulosyl-(1→4)-2-deoxy-α-L-Fucp-(1→4)-trifluoroacetyl-L-Rhodosamine

Syrup: [α]~D~ -202; Ref.:: 406

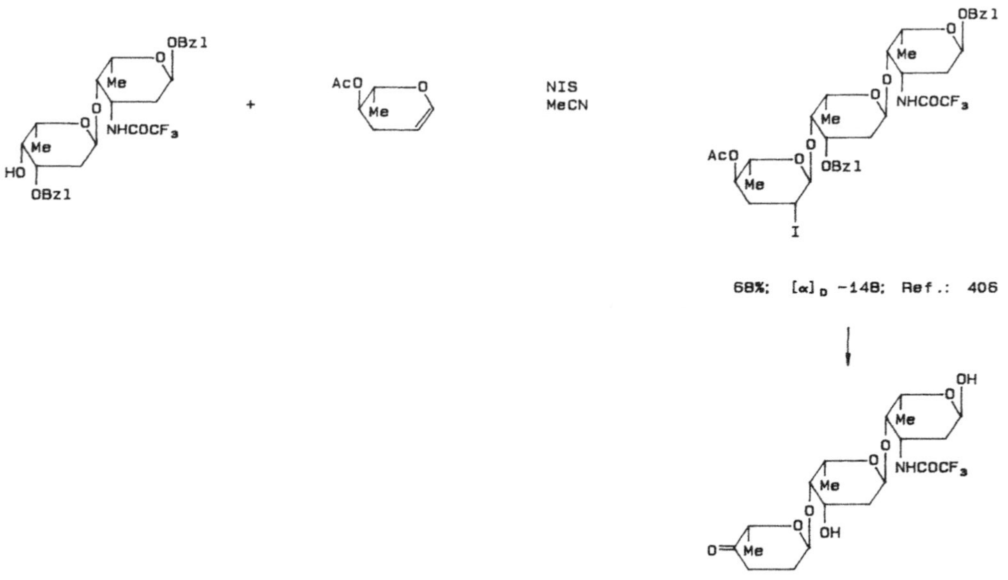

68%; [α]~D~ -148; Ref.:: 406

β-D-Galp-(1→4)-α-D-GlcpNAc-(1→6)-7-deoxy-D-glycero-D-galacto-Hepp

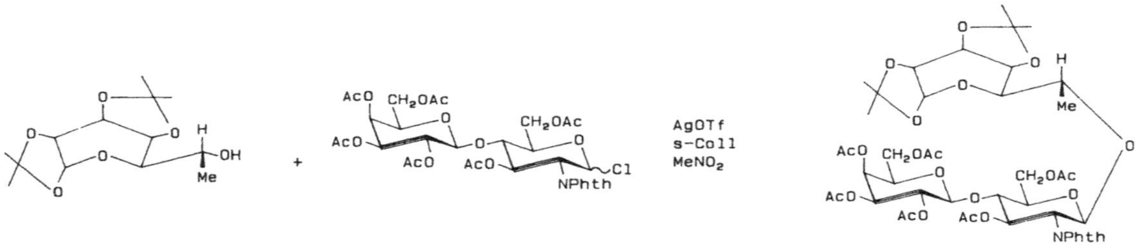

AgOTf
s-Coll
MeNO₂

65%; m.p. 114—119°C; [α]ᴅ −20 (chloroform); Ref.: 396

β-D-Galp-(1→4)-α-D-GlcpNAc-(1→6)-7-deoxy-L-glycero-D-galacto-Hepp

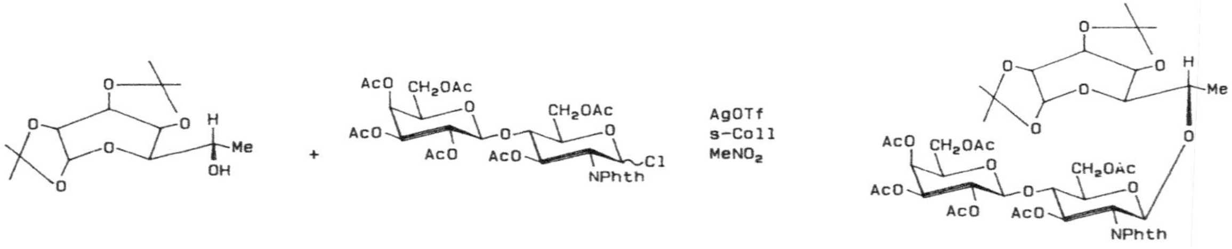

AgOTf
s-Coll
MeNO₂

36%; m.p. 119—123°C; [α]ᴅ −21.6 (chloroform); Ref.: 396

REFERENCES

1. **Schmidt, R. R. and Hermentin, P.**, α-Connected di- and tri-saccharides of D-ribofuranose, *Chem. Ber.*, 112, 2659, 1979.
2. **Schmidt, R. R. and Hermentin, P.**, α-Glycosidic linkage of D-ribofuranose, *Angew. Chem.*, 89, 58, 1977.
3. **Kosma, P., Christian, R., Schulz, G., and Unger, F. M.**, Synthesis of alternate linear and branched repeating units of the *Escherichia coli* LP 1092 capsular polysaccharide containing 3-deoxy-α-D-*manno*-2-octulosonic acid (KDO) linked to secondary positions of D-ribose, *Carbohydr. Res.*, 141, 239, 1985.
4. **Hatanaka, K. and Kuzuhara, H.**, A general method for stepwise elongation of the (1→5)-α-D-arabinofuranan chain, *J. Carbohydr. Chem.*, 4, 333, 1985.
5. **Lipták, A., Szurmai, Z., Nánási, P., and Neszmélyi, A.**, ¹³C-NMR Study of methyl- and benzyl-ethers of L-arabinose and oligosaccharides having L-arabinose at the reducing end. Synthesis of 2-*O*-β-D-glucopyranosyl-, 2-*O*-α-L-rhamnopyranosyl-, 3-*O*-β-D-glucopyranosyl-2-*O*-α-L-rhamnopyranosyl- and 4-*O*-β-D-glucopyranosyl-2-*O*-α-L-rhamnopyranosyl-L-arabinose, *Tetrahedron*, 38, 3489, 1982.
6. **Hirsch, J. and Kováč, P.**, Chemical synthesis of xylotriose (4-*O*-(4-*O*-β-D-xylopyranosyl-β-D-xylopyranosyl)-D-xylopyranose), *Chem. Zvesti*, 36, 125, 1982.
7. **Hirsch, J. and Petráková, E.**, Sequential synthesis and ¹³C NMR spectra of methyl 3-*O*- and 2-*O*-(β-D-xylobiosyl)-β-D-xylopyranosides, *Chem. Zvesti*, 38, 409, 1984.
8. **Dupeyre, D., Excoffier, G., and Utille, J.-P.**, Stepwise synthesis of linear β-D-(1→3)-xylo-oligosaccharides. Preparation of a β,β,β-D-linked tetrasaccharide derivative, *Carbohydr. Res.*, 135, C1, 1984.
9. **Koto, S., Morishima, N., Takenaka, K., Uchida, C., and Zen, S.**, Pentoside synthesis by dehydrative glycosylation. Synthesis of *O*-α-L-arabinofuranosyl-(1→3)-*O*-β-D-xylopyranosyl-(1→4)-D-xylopyranose, *Bull. Chem. Soc. Jpn.*, 58, 1464, 1985.
10. **Hirsch, J., Petráková, E., and Schraml, J.**, Stereoselective synthesis and ¹³C n.m.r. spectra of two isomeric methyl β-glycosides of trisaccharides related to arabinoxylan, *Carbohydr. Res.*, 131, 219, 1984.
11. **Kováč, P. and Hirsch, J.**, Stepwise synthesis of methyl 4-*O*-[3-*O*-(β-D-xylopyranosyl)-β-D-xylopyranosyl]-β-D-xylopyranoside, *Carbohydr. Res.*, 79, 303, 1980.
12. **Kováč, P. and Hirsch, J.**, Sequential synthesis and ¹³C-n.m.r. spectra of methyl β-glycosides of (1→4)-β-D-xylo-oligosaccharides, *Carbohydr. Res.*, 100, 177, 1982.
13. **Hirsch, J., Kováč, P., and Petráková, E.**, An approach to the systematic synthesis of (1→4)-β-D-xylo-oligosaccharides, *Carbohydr. Res.*, 106, 203, 1982.
14. **Hirsch, J. and Kováč, P.**, Synthesis of two isomeric methyl β-D-xylotriosides containing a (1→2)-β-linkage, *Carbohydr. Res.*, 77, 241, 1979.
15. **Kováč, P.**, Alternative syntheses of methylated sugars. XXII. Synthesis of methyl β-xylotrioside, *Chem. Zvesti*, 34, 234, 1980.
16. **Kováč, P. and Hirsch, J.**, Systematic, sequential synthesis of (1→4)-β-D-xylo-oligosaccharides and their methyl β-glycosides, *Carbohydr. Res.*, 90, C5, 1981.
17. **Hirsch, J., Kováč, P., Alföldi, J., and Mihálov, V.**, Synthesis of an aldotriouronic acid and derivative related to (4-*O*-methylglucurono)xylans, *Carbohydr. Res.*, 88, 146, 1981.
18. **Garegg, P. J., Lindberg, B., and Norberg, T.**, Synthesis of *O*-β-D-galactopyranosyl-(1→3)-*O*-β-D-galactopyranosyl-(1→4)-*O*-β-D-xylopyranosyl-(1→3)-L-serine, *Acta Chem. Scand. B*, 33, 449, 1979.
19. **Lindberg, B., Rodén, L., and Silvander, B.-G.**, Synthesis of oligosaccharides containing galactose and xylose, *Carbohydr. Res.*, 2, 413, 1966.
20. **Petráková, E. and Kováč, P.**, Synthesis of 2,4-di-*O*-β-D-xylopyranosyl-D-xylopyranose, *Chem. Zvesti*, 35, 699, 1981.
21. **Kováčik, V., Mihálov, V., and Kováč, P.**, Structural analysis by mass spectrometry of oligosaccharides related to xylans, *Carbohydr. Res.*, 88, 189, 1981.
22. **Kováč, P.**, Synthesis of methyl 3,4-di-*O*-(β-D-xylopyranosyl)-β-D-xylopyranoside,3,4-di-*O*-(β-D-xylopyranosyl)-β-D-xylopyranoside,3,4-di-*O*-(β-D-xylopyranosyl)-β-D-xylopyranoside, a methyl β-xylotrioside related to branched xylans, *Collect. Czech. Chem. Commun.*, 44, 928, 1979.
Chem. Commun., 44, 928, 1979.
23. **Kováč, P.**, Synthesis and reactions of uronic acid derivatives. Part XVIII. The stepwise synthesis of a (4-*O*-methyl-glucurono)xylan type branched aldotriuronic acid derivative, *J. Carbohydr. Nucleosides Nucleotides*, 4, 165, 1977.
24. **Jones, J. K. N. and Reid, P. E.**, The synthesis of 5-*O*-β-D-glucopyranosyl-D-xylose and 3,5-di-*O*-β-D-glucopyranosyl-D-xylose, *Can. J. Chem.*, 41, 2382, 1963.
25. **Koto, S., Yago, K., Zen, S., Tomonaga, F., and Shimada, S.**, α-D-Glucosylation by 6-*O*-acetyl-2,3,4-tri-*O*-benzyl-D-glucopyranose using trimethylsilyl triflate and pyridine. Synthesis of α-maltosyl and α-isomaltosyl α-D-glucosides, *Bull. Chem. Soc. Jpn.*, 59, 411, 1986.
26. **Koeppen, B. H.**, Synthesis of centose and two isomeric D-glucose trisaccharides, *Carbohydr. Res.*, 13, 417, 1970.

27. **Excoffier, G., Paillet, M., and Vignon, M.,** Cyclic(1→6)-β-D-glucopyranose oligomers: synthesis of cyclogentiotriose and cyclogentiotetraose peracetates, *Carbohydr. Res.*, 135, C10, 1985.

28. **Takeo, K.,** Synthesis of kojitriose, *Carbohydr. Res.*, 88, 158, 1981.

29. **Fujiwara, T., Takeda, T., and Ogihara, K.,** Synthesis of trisaccharides related to an arabinoglucan, *Carbohydr. Res.*, 141, 168, 1985.

30. **Wolfrom, M. L. and Koizumi, K.,** A chemical synthesis of panose, *Chem. Commun.*, 2, 1966.

31. **Wolfrom, M. L. and Koizumi, K.,** A chemical synthesis of panose and an isomeric trisaccharide, *J. Org. Chem.*, 32, 656, 1967.

32. **Heiferich, B. and Müller, W. M.,** The synthesis of α-D-glucopyranosyl derivatives, *Chem. Ber.*, 106, 2508, 1973.

33. **Koto, S., Morishima, N., Kihara, Y., Suzuki, H., Kosugi, S., and Zen, S.,** The stereoselective dehydrative α-glucosylation using 6-O-acetyl- and 6-O-p-nitrobenzoyl-2,3,4-tri-O-benzyl-D-glucopyranoses, *Bull. Chem. Soc. Jpn.*, 56, 188, 1983.

34. **Takeo, K. and Shinmitsu, K.,** A convenient synthesis of 6'-C-substituted β-maltose heptaacetates and of panose, *Carbohydr. Res.*, 133, 135, 1984.

35. **Gi, C. T. and Tejima, S.,** Synthesis of O-α-and O-β-D-galactopyranosyl-(1→6)-O-α-D-glucopyranosyl-(1→4)-D-glucopyranoses, *Chem. Pharm. Bull.*, 25, 464, 1977.

36. **Laesecke, K. and Schmidt, R. R.,** α-Glucosidase inhibitors. 2. Synthesis of modified maltotrioses, *Liebigs Ann. Chem.*, 1910, 1983.

37. **Pozsgay, V., Nánási, P., and Neszmélyi, A.,** Utilisation of the D-glucopyranosyl group as a non-participating group in stereoselective glycosylation: synthesis of O-α-D-glucopyranosyl-(1→2)-O-α-D-glucopyranosyl-(1→6)-D-glucose, *Carbohydr. Res.*, 75, 310, 1979.

38. **Morishima, N., Koto, S., Irisawa, T., Hashimoto, Y., Yamazaki, M., Higuchi, T., and Zen, S.,** Stereoselectivity in the dehydrative glucosylation with hepta-O-benzyl-glucobioses, *Chem. Lett.*, 1383, 1982.

39. **Takiura, K., Kakehi, K., and Honda, S.,** Studies of oligosaccharides. X. Synthesis of isomaltose and isomaltotriose by benzyl blocking method, *Chem. Pharm. Bull.*, 21, 523, 1973.

40. **Eby, R. and Schuerch, C.,** The stepwise synthesis of methyl α-isomaltooligoside derivatives and methyl α-isomaltopentaoside, *Carbohydr. Res.*, 50, 203, 1976.

41. **Eby, R. and Schuerch, C.,** A chemical synthesis of benzylated methyl α-isomalto-oligosaccharides, *Macromolecules*, 7, 397, 1974.

42. **Eby, R.,** The synthesis of α-isomalto-oligosaccharide derivatives and their protein conjugates, *Carbohydr. Res.*, 70, 75, 1979.

43. **Ogawa, T. and Horisaki, T.,** Synthesis of 2-O-hexadecanoyl-1-O-hexadecyl-[α-Glc-6SO₃Na-(1→6)-α-Glc-(1→6)-α-Glc-(1→3)]-sn-glycerol: a proposed structure for the glyceroglucolipids of human gastric secretion and of the mucous barrier of rat-stomach antrium, *Carbohydr. Res.*, 123, C1, 1983.

44. **Koto, S., Morishima, N., Irisawa, T., Hashimoto, Y., Yamazaki, M., and Zen, S.,** Biosylation of benzyl 2,3,4-tri-O-benzyl-α-D-glucopyranoside using p-nitrobenzenesulfonyl chloride, silver trifluoromethanesulfonate, and triethylamine. Synthesis of linear trisaccharides, *Nippon Kagaku Kaishi*, 1651, 1982.

45. **Ogawa, T., Nakabayashi, S., and Shibata, S.,** Synthetic studies on Nephritogenic glycosides. Synthesis of β-Glc-(1→6)-α-Glc-(1→6)-α-Glc-(1→Asn), *Agric. Biol. Chem.*, 47, 1353, 1983.

46. **Ogawa, T. and Takanashi, Y.,** Synthesis of β-D-(1→2)-linked-D-glucopentaose, a part of the structure of the exocellular β-D-glucan of *Agrobacterium tumefaciens*, *Carbohydr. Res.*, 123, C16, 1983.

47. **Takeo, K. and Suzuki, Y.,** Synthesis of the tri- and tetra-saccharides related to the fine structures of lichenan and cereal β-D-glucans, *Carbohydr. Res.*, 147, 265, 1986.

48. **Sakai, J., Takeda, T., and Ogihara, Y.,** Selective acetolysis of benzyl ethers of methyl D-glucopyranosides, *Carbohydr. Res.*, 95, 125, 1981.

49. **Nicolaou, K. C., Randall, J. L., and Furst, G. T.,** Stereospecific synthesis of Rhyncosporosides: a family of fungal metabolites causing scald disease in barley and other grasses, *J. Am. Chem. Soc.*, 107, 5556, 1985.

50. **Takeo, K., Okushio, K., Fukuyama, K., and Kuge, T.,** Synthesis of cellobiose, cellotriose, cellotetraose, and lactose, *Carbohydr. Res.*, 121, 163, 1983.

51. **Hall, D. M. and Lawler, T. E.,** New routes to the synthesis of 2,3,6-tri-O-substituted methyl β-D-glucopyranosides. An improved synthesis of α-cellotriose hendecaacetate, *Carbohydr. Res.*, 16, 1, 1971.

52. **Takeo, K., Yasato, T., and Kuge, T.,** Synthesis of α- and β-cellotriose hendecaacetates and of several 6,6',6''-tri-substituted derivatives of methyl β-cellotrioside, *Carbohydr. Res.*, 93, 148, 1981.

53. **Sugawara, F., Nakayama, H., and Ogawa, T.,** Synthetic studies on rhynchosporoside: stereoselective synthesis of 1-O- and 2-O-α-cellotriosyl-3-deoxy-2(R)-and-2(S)-glycerol, *Carbohydr. Res.*, 123, C25, 1983.

54. **Sugawara, F., Nakayama, H., Strobel, G. A., and Ogawa, T.,** Stereoselective synthesis of 1- and 2-O-α-D-cellotriosyl-3-deoxy-2(R)- and 2(S)-glycerols related to rynchosporoside, *Agric. Biol. Chem.*, 50, 2261, 1986.

55. **Schmidt, R. R. and Michel, J.**, Synthese von linearen und verzweigten Cellotetraosen, *Angew. Chem.*, 94, 77, 1982.

56. **Roy, N. and Timell, T. E.**, The acid hydrolysis of glycosides. VIII. Synthesis and hydrolysis of three aldotriouronic acids, *Carbohydr. Res.*, 6, 475, 1968.

57. **Sato, S., Mori, M., Ito, Y., and Ogawa, T.**, An efficient approach to O-glycosides through CuBr₂-Bu₄NBr mediated activation of glycosides, *Carbohydr. Res.*, 155, C6, 1986.

58. **Koto, S., Morishima, N., Sato, H., Sato, Y., and Zen, S.**, Synthesis of O-β-D-glucopyranosyl-(1→2)-O-β-D-glucopyranosyl-(1→6)-D-glucopyranose, dehydrative β-D-glucosylation using 2-O-acetyl-3,4,6-tri-O-benzyl-D-glucopyranose, *Bull. Chem. Soc. Jpn.*, 58, 120, 1985.

59. **Kochetkov, N. K., Khorlin, A.Ya., Bochkov, A. F., Detushkina, L. B., and Zolotuhin, I. O.**, Synthesis of trisaccharides with the orthoester method, *Zhur. Obsch. Khim.*, 37, 1272, 1967.

60. **Kochetkov, N. K., Khorlin. A. Ya., and Bochkov, A. F.**, A new method of glycosylation, *Tetrahedron*, 23, 693, 1967.

61. **Kochetkov, N. K.**, A new synthesis of glycosides, *Kem. Kozl.*, 28, 425, 1967.

62. **Nánási, P., Lipták, A., and Jánossy, L.**, Synthesis of O-α-D-glucopyranosyl-(1→4)-O-β-D-glucopyranosyl-(1→6)-D-glucose, *Acta Chim. Acad. Sci. Hung.*, 88, 155, 1976.

63. **Ogawa, T., Nakabayashi, S., and Shibata, S.**, Synthetic studies on nephritogenic glycosides. Synthesis of α-Glc-(1→6)-β-Glc-(1→6)-α-Glc-(1→Asn), *Carbohydr. Res.*, 86, C7, 1980.

64. **Takeda, T., Sugiura, Y., Hamada, C., Fujii, R., Suzuki, K., Ogihara, Y., and Shibata, S.**, The nephritogenic glycopeptide from rat glomerular basement membrane. II. Synthesis of O-(α-D-glucopyranosyl)-(1→6)-O-β-D-glucopyranosyl-(1→6)-N-(L-β-aspartyl)-α-D-glucopyranosylamine (α-D-Glc-(1→6)-β-D-Glc-(1→6)-α-D-Glc-(1→Asn)), *Chem. Pharm. Bull.*, 29, 3196, 1981.

65. **Ogawa, T., Nakabayashi, S., and Shibata, S.**, Synthetic studies on cell surface glycans. Part XIX. Synthetic studies on nephritogenic glycosides. Synthesis of α-Glc-(1→6)-β-Glc-(1→6)-α-Glc-(1→Asn), *Agric. Biol. Chem.*, 47, 1213, 1983.

66. **Rychener, M., Bigler, P., and Pfander, H.**, Synthese und ¹H-NMR-Studie der vier unverzweigten peracetylierten β-D-glucopyranosyl-β-gentiobiosen, *Helv. Chim. Acta*, 67, 378, 1984.

67. **Ossowski, P., Pilotti, A., Garegg, P. J., and Lindberg, B.**, Syntheses of a branched hepta and an octasaccharide with phytoalexin-elicitor activity, *Angew. Chem.*, 95, 809, 1983.

68. **Ossowski, P., Pilotti, A., Garegg, P. J. and Lindberg, B.**, Synthesis of a glucoheptaose and a glucooctaose that elicit phytoalexin accumulation in soybean, *J. Biol. Chem.*, 259, 11337, 1984.

69. **Hall, D. M., Lawler, T. E., and Childress, B. C.**, A practical synthesis of 1,2,3,6-tetra-O-acetyl-α-and β-D-glucopyranose, and their use to prepare trisaccharides, *Carbohydr. Res.*, 38, 359, 1974.

70. **Takiura, K., Honda, S., Endo, T., and Kakehi, K.**, Studies of oligosaccharides. IX. Synthesis of gentiooligosaccharides by block condensation, *Chem. Pharm. Bull.*, 20, 438, 1972.

71. **Takiura, K., Yamamoto, M., Miyagi, Y., Takai, H., Honda, S., and Yuki, H.**, Studies of oligosaccharides. XV. Synthesis of hydroquinone glycosides of gentio-oligosaccharides, *Chem. Pharm. Bull.*, 22, 2451, 1974.

72. **Excoffier, G., Gagnaire, D. Y., and Vignon, M. R.**, Le groupe trichloroacétyle comme substituant temporaire; synthèse du gentiotétraose, *Carbohydr. Res.*, 46, 201, 1976.

73. **Nesmeyanov, V. A., Zurabyan, S. E., and Khorlin, A. Ya.**, Sugar acetates as glycosylating agents in oligosaccharide synthesis, *Tetrahedron Lett.*, 3213, 1973.

74. **Takabe, S., Takeda, T., and Ogihara, Y.**, Synthesis of glycosyl esters of oleanolic acid, *Carbohydr. Res.*, 76, 101, 1979.

75. **Neszmélyi, A., Lipták, A., Nánási, P., and Szejti, J.**, C-13 relaxation time gradients in complexes of linear oligosaccharides and cyclodextrin. A potential new method for sugar sequence determination, *Acta Chim. Hung.*, 113, 431, 1983.

76. **Shapiro, D.**, Partial synthesis of the carbohydrate chain of brain gangliosides, *Chem. Phys. Lipids*, 5, 80, 1970.

77. **Antonenko, T. S., Zurabyan, S. E., and Khorlin, A. Ya.**, Synthesis of oligosaccharides having (1→6)-1,2-trans-N-acetylglucoseamidinic linkage, *Izv. Akad. Nauk. SSSR Ser. Khim.*, 2766, 1970.

78. **van Boeckel, C. A. A., Westerduin, P., and van Boom, J. H.**, Synthesis of two purple-membrane glycolipids and the glycolipid sulfate O-(β-D-glucopyranosyl 3-sulfate)-(1→6)-O-α-D-mannopyranosyl-(1→2)-O-α-D-glucopyranosyl-(1→1)-2,3-di-O-phytanyl-sn-glycerol, *Charbohydr. Res.*, 133, 219, 1984.

79. **van Boeckel, C. A. A., Westerduin, P., and Van Boom, J. H.**, Synthesis of 2,3-di-O-phytanyl-1-O-[glucosyl(galactosyl)-β(1→6)-mannosyl-α(1→2)-glucosyl-α(1→1)]-sn-glycerol. Purple membrane glycolipids, *Tetrahedron Lett.*, 22, 2819, 1981.

80. **Verez Bencomo, V., Garcia Fernandez, G., Basterrechea Rey, M., Coll Manchado, F., and Rodriquez Fernandez, M. E.**, Synthesis of methyl 3-0-(2-0-α-D-mannopyranosyl-α-D-mannopyranosyl)-α-D-glucopyranoside, *Rev. Cubana Farm.*, 17, 36, 1983.

81. **Torgov, V. I., Kudashova, O. V., Shibaev, V. N., and Kochetkov, N. K.**, The synthesis of analogs of Salmonella O-antigenic polysaccharide repeating unit fragments: rhamnopyranosyl-(α1→3)-glucose and mannopyranosyl-(α1→4)-rhamnopyranosyl-(α1→3)glucose, *Bioorg. Khim.*, 8, 114, 1982.

82. **Lafitte, C., Nguyen Phuoc Du, A.-M., Winternitz, F., Wylde, R., and Pratviel-Sosa, F.**, Synthése et étude R. M. N. de disaccharides et trisaccharides dans la série du L-rhamnose, *Carbohydr. Res.*, 67, 91, 1978.

83. **Doboszewski, B. and Zamojski, A.**, The synthesis of 2-O-α-D-galactopyranosyl-D-galactopyranose and 2-O-(2-O-α-D-galactopyranosyl-α-D-galactopyranosyl)-D-glucopyranose undeca-acetate, *Carbohydr. Res.*, 132, 29, 1984.

84. **Norberg, T. and Ritzen, H.**, Synthesis of methyl α-D-glucopyranosyl-(1→2)-α-D-galactopyranosyl-(1→3)-α-D-glucopyranoside and an acyclic analogue thereof for probing the carbohydrate-binding specificity of bacteriophage φX174, *Glycoconjugate J.*, 3, 135, 1986.

85. **Dahmén, J., Frejd, T., Magnusson, G., Noori, G., and Carlström, A.-S.**, Synthesis of spacer-arm, lipid, and ethyl glycosides of the trisaccharide portion [α-D-Gal-(1→4)-β-D-Gal-(1→4)-β-D-Glc] of the blood-group P^k antigen: preparation of neoglycoproteins, *Carbohydr. Res.*, 127, 15, 1984.

86. **Adachi, R. and Suami, T.**, Synthesis of manninotriose undecaacetate, *Bull. Chem. Soc. Jpn.*, 50, 1901, 1977.

87. **Shapiro, D. and Acher, A. J.**, Total synthesis of ceramide trihexoside accumulating with Fabry's disease, *Chem. Phys. Lipids*, 22, 197, 1978.

88. **Shapiro, D., Acher, A. J., Robinsohn, Y., and Diver-Haber, A.**, Studies in the ganglioside series. VI. Synthesis of the trisaccharide inherent in Tay-Sachs ganglioside, *J. Org. Chem.*, 36, 832, 1971.

89. **Batavyal, L. and Roy, N.**, Synthesis of O-(β-D-glucopyranosyluronic acid)-(1→3)-O-β-D-galactopyranosyl-(1→4)-D-glucopyranose, *Carbohydr. Res.*, 145, 328, 1986.

90. **Paulsen, H., Hadamczyk, D., Kutschker, W., and Bünsch, A.**, Regioselektive Glycosylierung an 3'-OH oder 4'-OH der Lactose durch Einsatz unterschiedlicher Katalysator-Systeme, *Liebigs Ann. Chem.*, 129, 1985.

91. **Acher, A. J., Robinsohn, Y., Rachaman, E. S., and Shapiro, D.**, Studies in the ganglioside series. V. Synthesis of 2-acetamido-2-deoxy-0-β-D-glucopyranosyl-(1→3)-O-β-D-galactopyranosyl-(1→4)-D-glucose, *J. Org. Chem.*, 35, 2436, 1970.

92. **Maranduba, A. and Veyrières, A.**, Glycosylation of lactose. Synthesis of methyl O-(2-acetamido-2-deoxy-β-D-glucopyranosyl)-(1→3)-O-β-D-galactopyranosyl-(1→4)-β-D-glucopyranoside and methyl O-β-D-galactopyranosyl-(1→4)-O-(2-acetamido-2-deoxy-β-D-glucopyranosyl)-(1→3)-O-β-D-galactopyranosyl-(1→4)-β-D-glucopyranoside, *Carbohydr. Res.*, 135, 330, 1985.

93. **Paulsen, H., Paal, M., Hadamczyk, D., and Steiger, K.-M.**, Regeoselektive Glycosidierung von Lactose durch unterschiedliche Katalysatorsysteme. Synthese der Saccharid-Sequenzen von asialo-G$_{M1}$ und asialo-G$_{M2}$-Gangliosiden, *Carbohydr. Res.*, 131, C1, 1984.

94. **Ito, Y., Sugimoto, M., Sato, S., and Ogawa, T.**, Total synthesis of a lacto-ganglio series glycosphingolipid, *Tetrahedron Lett.*, 4753, 1986.

95. **Maranduba, A. and Veyriéres, A.**, Glycosylation of lactose: synthesis of branched oligosaccharides involved in the biosynthesis of glycolipids having blood-group I activity, *Carbohydr. Res.*, 151, 105, 1986.

96. **Sato, S., Ito, Y., and Ogawa, T.**, Stereo- and regio-controlled total synthesis of the Le^b antigen, III^{4}FucIV^{2}FucLcOse$_4$Cer, *Carbohydr. Res.*, 155, C1, 1986.

97. **Wessel, H.-P., Iversen, T., and Bundle, D. R.**, Synthesis of the trisaccharide moiety of gangliotriosyl-ceramide (asialo GM2), *Carbohydr. Res.*, 130, 5, 1984.

98. **Zurabyan, S. E., Markin, V. A., Pimenova, V. V., Rozynov, B. V., Sadovskaya, V. L., and Khorlin, A. Ya.**, Synthesis of tri- and tetrasaccharides, structural isomers of milk oligosaccharides, *Bioorg. Khim.*, 4, 928, 1978.

99. **Matsuda, H., Ishihara, H., and Tejima, S.**, Chemical modification of lactose. XII. Preparation of O-(2-acetamido-2-deoxy-β-D-glucopyranosyl)-(1→6)-O-β-D-galactopyranosyl-(1→4)-D-glucopyranose (6'-N-acetylglucosaminyllactose), *Chem. Pharm. Bull.*, 27, 2564, 1979.

100. **Takamura, T., Chiba, T., and Tejima, S.**, Chemical modification of lactose. XIV. Synthesis of O-2-acetamido-2-deoxy-β-D-glucopyranosyl-(1→3)-O-[2-acetamido-2-deoxy-β-D-glucopyranosyl-(1→6)]-O-β-D-galactopyranosyl-(1→4)-β-D-glucopyranose (3',6'-di-β-N-acetylglucosaminyl-β-lactose), *Chem. Pharm. Bull.*, 29, 1027, 1981.

101. **Jacquinet, J.-C. and Sinaÿ, P.**, Chemical synthesis of the human P^k-antigenic determinant, *Carbohydr. Res.*, 143, 143, 1985.

102. **Garegg, P. J. and Hultberg, H.**, Synthesis of di- and trisaccharides corresponding to receptor structures recognised by pyelonephritogenic E. coli fimbriae (pili), *Carbohydr. Res.*, 110, 261, 1982.

103. **Cox, D. D., Metzner, E. K., and Reist, E. J.**, The synthesis of methyl 4-0-(4-0-α-D-galactopyranosyl-β-D-galactopyranosyl)-β-D-glucopyranoside: the methyl β-glycoside of the trisaccharide related to Fabry's disease, *Carbohydr. Res.*, 63, 139, 1978.

104. **Paulsen, H. and Bünsch, A.**, Synthese der Pentasaccharid-Kette des Forssman-antigens, *Carbohydr. Res.*, 100, 143, 1982.

105. **Beith-Halahmi, D., Flowers, H. M., and Shapiro, D.**, Synthesis of *O*-β-D-galactopyranosyl-(1→3)-*O*-β-D-galactopyranosyl-(1→4)-D-glucose, *Carbohydr. Res.*, 5, 25, 1967.

106. **Beith-Halahmi, D. and Flowers, H. M.**, Substituted cerebrosides. Part V. Synthesis of a dihydroceramide trihexoside, *Carbohydr. Res.*, 8, 340, 1968.

107. **Chung, T. G., Ishihara, H., and Tejima, S.**, Chemical modification of lactose. IX. Synthesis of *O*-β-D-galactopyranosyl-(1→6)-*O*-β-D-galactopyranosyl-(1→4)-D-glucopyranose (6'-galactosyl-lactose), *Chem. Pharm. Bull.*, 26, 2147, 1978.

108. **Paulsen, H. and Paal, M.**, Synthese der Tetra- und Trisaccharidsequenzen von Asialo-G$_{M1}$ und -G$_{M2}$. Lenkung der Regioselektivität der Glycosidierung von Lactose, *Carbohydr. Res.*, 137, 39, 1985.

109. **Sebesan, S. and Lemieux, R. U.**, Synthesis of tri- and tetrasaccharide haptens related to the *Asialo* forms of the gangliosides G$_{M2}$ and G$_{M1}$, *Can. J. Chem.*, 62, 644, 1984.

110. **Sugimoto, M., Horisaki, T., and Ogawa, T.**, Synthetic studies on cell-surface glycans. Part 35. Stereoselective synthesis of asialo-G$_{M1}$- and asialo G$_{M2}$-ganglioside, *Glycoconjugate J.*, 2, 11, 1985.

111. **Shapiro, D., Acher, A. J., and Robinsohn, Y.**, Studies in the ganglioside series. VII. Total synthesis of Tay-Sachs' globoside, *Chem. Phys. Lipids*, 10, 28, 1973.

112. **Abbas, S. A., Barlow, J. J., and Matta, K. L.**, Synthesis of *O*-α-L-fucopyranosyl-(1→2)-*O*-β-D-galactopyranosyl-(1→4)-D-glucopyranose (2'-*O*-α-L-fucopyranosyl-lactose), *Carbohydr. Res.*, 88, 51, 1981.

113. **Fernandez-Mayoralas, A., and Martin-Lomas, M.**, Synthesis of 3- and 2'-fucosyl-lactose and 3,2'-difucosyl-lactose from partially benzylated lactose derivatives, *Carbohydr. Res.*, 154, 93, 1986.

114. **Baer, H. H. and Abbas, S. A.**, Synthesis of *O*-α-L-fucopyranosyl-(1→3)-*O*-β-D-galactopyranosyl-(1→4)-D-glucose (3'-*O*-α-L-fucopyranosyllactose), and an improved route to its β-(1″-3′)-linked isomer, *Carbohydr. Res.*, 84, 53, 1980.

115. **Takamura, T., Chiba, T., and Tejima, S.**, Chemical modification of lactose. XV. Synthesis of *O*-α- and *O*-β-L-fucopyranosyl-(1→3)-*O*-β-D-galactopyranosyl-(1→4)-D-glucopyranoses (3'-O-α- and 3'-*O*-β-L-fucopyranosyllactose), *Chem. Pharm. Bull.*, 29, 1076, 1981.

116. **Chiba, T. and Tejima, S.**, Chemical modification of lactose. XVII. Synthesis of *O*-α- and *O*-β-L-fucopyranosyl-(1→4)- or -(1→6)-*O*-β-D-galactopyranosyl-(1→4)-D-glucopyranoses (4'- or 6'-*O*-α and -*O*-β-L-fucopyranosyllactose), *Chem. Pharm. Bull.*, 31, 75, 1983.

117. **Baer, H. H. and Abbas, S. A.**, Synthesis of *O*-α-L-fucopyranosyl-(1→6)-*O*-β-D-galactopyranosyl-(1→4)-D-glucopyranose (6'-*O*-α-L-fucopyranosyllactose), *Carbohydr. Res.*, 83, 146, 1980.

118. **Baer, H. H. and Abbas, S. A.**, Synthesis of *O*-β-L-fucopyranosyl-(1→3)-*O*-β-D-galactopyranosyl-(1→4)-D-glucopyranose (3'-*O*-β-L-fucopyranosyllactose), *Carbohydr. Res.*, 77, 117, 1979.

119. **Ogawa, T. and Sugimoto, M.**, Synthesis of α-Neu5Ac*p*-(2→3)-D-Gal and α-Neu5Ac*p*-(2→3)-β-D-Gal*p*-(1→4)-D-Glc, *Carbohydr. Res.*, 135, C5, 1985.

120. **Paulsen, H. and von Deessen, V.**, Glycosidsynthese von *N*-Acetylneuraminsäure mit sekundären Hydroxylgruppen, *Carbohydr. Res.*, 146, 147, 1986.

121. **Sugimoto, M. and Ogawa, T.**, Synthetic studies on cell-surface glycans. Part 34. Synthesis of a hematoside (G$_{M3}$-ganglioside) and a stereoisomer, *Glycoconjugate J.*, 2, 5, 1985.

122. **Okamoto, K., Kondo, T., and Goto, T.**, An effective synthesis of α-glycosides of *N*-acetylneuraminic acid by use of 2β-halo-3β-hydroxy-4,7,8,9-tetra-*O*-acetyl-*N*-acetylneuraminic acid methyl ester, *Tetrahedron Lett.*, 27, 5233, 1986.

123. **Furuhata, K., Anazawa, K., Itoh, M., Shitori, Y., and Ogura, H.**, Studies on sialic acids. V. Synthesis of α- and β-D-Neu5Ac*p*-(2→6)-lactose, *Chem. Pharm. Bull.*, 34, 2725, 1986.

124. **Paulsen, H. and Paal, M.**, Synthese der T-antigenen Trisaccharide *O*-β-D-Galactopyranosyl-(1→3)-*O*-(2-acetamido-2-desoxy-α-D-galactopyranosyl)-(1→6)-D-galactopyranose und *O*-β-D-Galactopyranosyl-(1→3)-*O*-(2-acetamido-2-desoxy-α-D-galactopyranosyl)-(1→6)-D-glucopyranose und deren Anknüpfung an Proteine, *Carbohydr. Res.*, 113, 203, 1983.

125. **Paulsen, H. and Paal, M.**, Blocksynthese von D-Glycopeptiden und anderen T-Antigen Strukturen, *Carbohydr. Res.*, 135, 71, 1984.

126. **Flowers, H. M.**, Synthesis of oligosaccharides of L-fucose containing α- and β-anomeric configurations in the same molecule, *Carbohydr. Res.*, 119, 75, 1983.

127. **Suami, T., Otake, T., Nishimura, T., and Ikeda, T.**, Synthesis of planteose, *Bull. Chem. Soc. Jpn.*, 46, 1014, 1973.

128. **Okamoto, K., Kondo, T., and Goto, T.**, A stereospecific synthesis of β-glycosides of *N*-acetylneuraminic acid and secondary alcohols, *Chem. Lett.*, 1449, 1986.

129. **Eby, R. and Schuerch, C.**, The synthesis of trisaccharide antigenic determinants for the branch points in natural dextrans and their protein conjugates, *Carbohydr. Res.*, 79, 53, 1980.

130. **Koto, S., Morishima, N., Kusuhara, C., Sekido, S., Yoshida, T., and Zen, S.**, Stereoselective α-glucosylation with tetra-*O*-benzyl-α-D-glucose and a mixture of trimethylsilyl bromide, cobalt (II) bromide, tetrabutylammonium bromide, and a molecular sieve. A synthesis of 3,6-di-*O*-(α-D-glucopyranosyl)-D-glucose, *Bull. Chem. Soc. Jpn.*, 55, 2995, 1982.

131. **Ogawa, T. and Kaburagi, T.**, Synthesis of a branched D-glucoheptaose: the repeating unit of extracellular α-D-glucan 1355-S of *Leuconostoc mesenteroids* NRRL B-1355, *Carbohydr. Res.*, 110, C12, 1982.

132. **Koto, S., Inada, S., Yoshida, T., Toyama, M., and Zen, S.**, One-stage glycosylation using protected glycose: the synthesis of *O*-β-D-glucopyranosyl-(1→3)-*O*-[β-D-glucopyranosyl-(1→6)]-D-glucopyranose, *Can. J. Chem.*, 59, 255, 1981.

133. **DeSouza, R. and Goldstein, I. J.**, The synthesis of 4,6-di-*O*-(α-D-glucopyranosyl)-D-glucopyranose, the branch point of glycogen, *Tetrahedron Lett.*, 1215, 1964.

134. **Morishima, N., Koto, S., and Zen, S.**, Dehydrative α-glycosylation using a mixture of *p*-nitrobenzene-sulfonyl chloride, silver trifluoromethanesulfonate, *N,N*-dimethylacetamide, and triethylamine, *Chem. Lett.*, 1039, 1982.

135. **Koto, S., Morishima, N., Owa, M., and Zen, S.**, A stereoselective α-glucosylation by use of a mixture of 4-nitrobenzenesulfonyl chloride, silver trifluoromethanesulfonate, *N,N*-dimethylacetamide, and triethylamine, *Carbohydr. Res.*, 130, 73, 1984.

136. **Bock, K. and Pedersen, H.**, Assignment of the NMR parameters of the branch-point trisaccharide of amylopectin using 2-D NMR spectroscopy at 500 MHz, *J. Carbohydr. Chem.*, 3, 581, 1984.

137. **Goldstein, I. J. and Lindberg, B.**, The synthesis and characterisation of 4-α, 6-β-*bis*-D-glucopyranosido-D-glucose, *Acta Chem. Scand.*, 16, 383, 1962.

138. **Chung, T. G. and Tejima, S.**, Syntheses of *O*-α- and *O*-β-D-galactopyranosyl-(1→6)-*O*-[α-D-glucopy-ranosyl-(1→4)]-D-glucopyranoses, *Chem. Pharm. Bull.*, 26, 3562, 1978.

139. **Morishima, N., Koto, S., Uchino, M., and Zen, S.**, Synthesis of a trifurcated tetrasaccharide using dehydrative glycosylation, *Chem. Lett.*, 1183, 1982.

140. **Takeo, K.**, Silver trifluoromethanesulfonate-promoted Koenigs-Knorr reaction of methyl 4,6-*O*-benzyli-dene-β-D-glucopyranoside with 2,3,4,6-tetra-*O*-acetyl-α-D-glucopyranosyl bromide, *Carbohydr. Res.*, 87, 147, 1980.

141. **Temeriusz, A., Piekarska, B., Radomski, J., and Stepinski, J.**, A novel observation on Koenigs-Knorr condensation of 2,3,4,6-tetra-*O*-acetyl-α-D-glucopyranosyl bromide with methyl 4,6-*O*-benzylidene-α-D-glucopyranoside, *Polish J. Chem.*, 56, 141, 1982.

142. **Klemer, A., Buhe, E., Gundlach, F., and Uhlemann, G.**, Synthesen und Abbau von Oligosacchariden, *Forschungsber. des Landes Nordrhein-Westfalen*, Nr. 2393, 1974.

143. **Klemer, A. and Homberg, K.**, Über die Struktur des verzweigten Trisaccharids aus β-Benzyl-4,6-benzal-D-glucosid und α-Acetobrom-D-glucose: 3,6-bis-[β-D-glucosido(1.5)]-D-glucose(1.5), *Chem. Ber.*, 94, 2747, 1961.

144. **Ogawa, T. and Kaburagi, T.**, Synthesis of a branched D-glucotetraose, the repeating unit of the extracellular polysaccharides of *Grifola umbellate*, *Sclerotinia libertiana*, *Porodisculus pendulus* and *Schizophyllum commune* Fries, *Carbohydr. Res.*, 103, 53, 1982.

145. **Takeo, K.**, The Koenigs-Knorr reaction of methyl 4,6-*O*-benzylidene-β-D-glucopyranoside with 2,3,4,6-tetra-*O*-acetyl-α-D-glucopyranosyl bromide, *Carbohydr. Res.*, 77, 131, 1979.

146. **Takeo, K. and Tei, S.**, Synthesis of the repeating unit of schizophyllan, *Carbohydr. Res.*, 145, 293, 1986.

147. **Sztaricskai, F., Lipták, A., Pelyvás, I. F., and Bognár, R.**, Structural investigation of the antibiotic ristomycin A. Synthesis of ristobiose and ristotriose, *J. Antibiotics*, 29, 626, 1976.

148. **Temeriusz, A., Piekarska, B., Radomski, J., and Stepinski, J.**, The Koenigs-Knorr reaction of methyl 4,6-*O*-benzylidene-α-D-glucopyranoside with 2,3,4,6-tetra-*O*-acetyl-α-D-galactopyranosyl bromide, *Carbohydr. Res.*, 108, 298, 1982.

149. **Iversen, T. and Bundle, D. R.**, Synthesis of the colitose determinant of *Escherichia coli* 0111 and 3,6-di-*O*-(α-D-galactopyranosyl)-α-D-glucopyranoside, *Can. J. Chem.*, 60, 299, 1982.

150. **Weidmann, H., Appenroth, M., Leipert-Klug, R., Dax, K., and Stökl, P.**, Reactions of D-glucuronic acid. II. Disaccharides from 1,2-*O*-alkylidene-α-D-glucofuranurono-6,3-lactone, *J. Carbohydr. Nucleosides Nucleotides*, 3, 235, 1976.

151. **Betaneli, V. I., Litvak, M. M., Struchkova, M. I., Backinowsky, L. V., and Kochetkov, N. K.**, Glycosylation, acylation, and tritylation of methyl 1,2-*O*-cyanoethylidene-α-D-glucopyranuronate. Preparation of monomers for the synthesis of homo- and heteropolyuronides, *Bioorg. Khim.*, 9, 87, 1983.

152. **Paulsen, H. and Stenzel, W.**, Stereoselektive Synthese α-glycosidish verknüpfter Di- und Oligosaccharide der 2-Amino-2-desoxy-D-glucopyranose, *Chem. Ber.*, 111, 2334, 1978.

153. **Jacquinet, J.-C., Petitou, M., Duchaussoy, P., Lederman, I., Choay, J., Torri, G., and Sinaÿ, P.**, Synthesis of heparin fragments. A chemical synthesis of the trisaccharide *O*-(2-deoxy-2-sulfamido-3,6-di-*O*-sulfo-α-D-glucopyranosyl)-(1→4)-*O*-(2-*O*-sulfo-α-L-idopyranosyl)-uronic acid)-(1→4)-2-deoxy-2-sul-famido-6-*O*-sulfo-D-glucopyranose heptasodium salt, *Charbohydr. Res.*, 130, 221, 1984.

154. **Paulsen, H., Stiem, M., und Unger, F. M.**, Synthese eines 3-Desoxy-D-manno-2-octulosonsäure(KDO)-haltigen Tetrasaccharides und dessen Strukturvergleich mit einem Abbau-Produkt aus Bakterien-Lipopo-lysacchariden, *Tetrahedron Lett.*, 27, 1135, 1986.

155. **Khorlin, A. Ya., Nesmeyanov, V. A., and Zurabyan, S. E.**, Glycosylation of sugar 2,3-diphenyl-2-cyclopropen-1-yl ethers. A new route to oligosaccharides, *Carbohydr. Res.*, 43, 69, 1975.

156. **Zurabyan, S. E., Kolomeer, G. G., and Khorlin, A. Ya.**, Synthesis of 3-*O*-(β-D-galactopyranosyl)-6-*O*-(α-L-fucopyranosyl)-, and 4-*O*-(α-L-fucopyranosyl)-*N*-acetyl-D-glucosamine by the diphenylcyclopropenyl method, *Bioorg. Khim.*, 4, 654, 1978.

157. **Hindsgaul, O., Norberg, T., LePendu, J., and Lemieux, R. U.**, Synthesis of type 2 human blood-group antigenic determinants. The H, X, and Y haptens and variations of the H type 2 determinant as probes for the combining site of the lectin I of *Ulex europaeus*, *Carbohydr. Res.*, 109, 109, 1982.

158. **Zurabyan, S. E., Nesmeyanov, V. A., and Khorlin, A. Ya.**, Synthesis of the branched tetrasaccharide 2-acetamido-4-*O*-β-D-galactopyranosyl-6-*O*-[*O*-β-D-galactopyranosyl-(1→4)-β-D-glucopyranosyl]-2-deoxy-D-glucose, *Izv. Akad. Nauk SSSR Ser. Khim.*, 1421, 1976.

159. **Amvam-Zollo, P.-H. and Sinaÿ, P.**, *Streptococcus pneumoniae* type XIV polysaccharide: synthesis of a repeating branced tetrasaccharide with dioxa-type spacer-arms, *Carbohydr. Res.*, 150, 199, 1986.

160. **Ichikawa, Y., Ichikawa, R., and Kuzuhara, H.**, Synthesis, from cellobiose, of a trisaccharide closely related to the GlcNAc → GlcA → GlcN segment of the antithrombin-binding sequence of heparin, *Carbohydr. Res.*, 141, 273, 1985.

161. **Paulsen, H. and Stenzel, W.**, Synthese α-1→4- und α-1→3-verknüpfter Disaccharide der 2-Amino-2-desoxy-D-glucopyranose nach der Azid-Methode, *Chem. Ber.*, 111, 2348, 1978.

162. **Paulsen, H. and Stenzel, W.**, Bausteine von Oligosacchariden. Synthese α-glykosidisch verknüpfter 2-Aminozucker-Oligosaccharide, *Angew. Chem.*, 87, 547, 1975.

163. **Kinzy, W. and Schmidt, R. R.**, Synthese des Trisaccharides aus der "Repeating Unit" des Kapselpolysaccharids von *Neisseria meningitidis* (Serogruppe L), *Liebigs Ann. Chem.*, 1537, 1985.

164. **Auge, C., Warren, C. D., Jeanloz, R. W., Kiso, M., and Anderson, L.**, The synthesis of *O*-β-D-mannopyranosyl-(1→4)-*O*-(2-acetamido-2-deoxy-β-D-glucopyranosyl)-(1→4)-2-acetamido-2-deoxy-D-glucopyranose. Part II, *Carbohydr. Res.*, 82, 85, 1980.

165. **Zurabyan, S. E., Pimenova, V. V., Shashkova, E. A., and Khorlin, A. Ya.**, Synthesis of modified inhibitors of lysosim by oxazoline method, *Khim. Prir. Soedin.*, 7, 689, 1971.

166. **Ogawa, T., Kitajima, T., and Nukada, T.**, Synthesis of a protected trihexosyl unit: a glycosyl acceptor corresponding to the core structure of the N-linked glycan of a glycoprotein, *Carbohydr. Res.*, 123, C5, 1983.

167. **Shaban, M. and Jeanloz, R. W.**, The synthesis of *O*-α-D-mannopyranosyl-(1→6)-*O*-(2-acetamido-2-deoxy-β-D-glucopyranosyl)-(1→4)-2-acetamido-2-deoxy-D-glucose, *Carbohydr. Res.*, 19, 311, 1971.

168. **Shaban, M. A. E. and Jeanloz, R. W.**, The synthesis of oligosaccharide-asparagine compounds. V. *O*-α-D-Mannopyranosyl-(1→6)-*O*-(2-acetamido-2-deoxy-β-D-glucopyranosyl)-(1→4)-2-acetamido-*N*-(L-aspart-4-oyl)-2-deoxy-β-D-glucopyranosylamine, *Carbohydr. Res.*, 26, 315, 1973.

169. **Warren, C. D., Augé, C., Laver, M. L., Suzuki, S., Power, D., and Jeanloz, R. W.**, The synthesis of *O*-β-D-mannopyranosyl-(1→4)-*O*-(2-acetamido-2-deoxy-β-D-glucopyranosyl)-(1→4)-2-acetamido-2-deoxy-D-glucopyranose. Part I, *Carbohydr. Res.*, 82, 71, 1980.

170. **Paulsen, H. and Lebuhn, R.**, Synthese der invarianten Pentasaccharid-Core-Region der Kohlenhydrat-Ketten der N-Glycoproteine, *Carbohydr. Res.*, 130, 85, 1984.

171. **Zurabyan, S. E., Antonenko, T. S., and Khorlin, A. Ya.**, Oxazoline synthesis of 1,2-*trans*-2-acetamido-2-deoxyglycosides. Glycosylation of secondary hydroxyl groups in partially protected saccharides, *Carbohydr. Res.*, 15, 21, 1970.

172. **Zurabyan, S. E., Volosyuk, T. P., and Khorlin, A. J.**, Oxazoline synthesis of 1,2-*trans*-2-acetamido-2-deoxyglycosides, *Carbohydr. Res.*, 9, 215, 1969.

173. **Westerduin, P., van Boom, J. H., van Boeckel, C. A. A., and Beetz, T.**, Synthesis of two analogues of *Rhodomicrobium vannielii* lipid A, *Carbohydr. Res.*, 137, C4, 1985.

174. **Paulsen, H., Hayauchi, Y., and Unger, F. M.**, Bausteine von Oligosacchariden, LIII. Synthese von 3-Desoxy-D-*manno*-2-octulosonsäure-(KDO)-haltigen Trisacchariden, *Liebigs Ann. Chem.*, 1288, 1984.

175. **Paulsen, H., Heume, M., Györgydeák, Z., and Lebuhn, R.**, Synthese einer verzweigten Pentasaccharid-Sequenz der "bisected" Struktur von N-Glycoproteinen, *Carbohydr. Res.*, 144, 57, 1985.

176. **Itoh, Y. and Tejima, S.**, Synthesis of acetylated trisaccharides, Manα(1→3)-Manβ(1→4)GlcNAc and Manα(1→2)Manβ(1→4)GlcNAc, relating to mannosidosis, *Chem. Pharm. Bull.*, 31, 1632, 1983.

177. **Paulsen, H. and Lebuhn, R.**, Bausteine von Oligosacchariden, XLVII. Synthese von Tri- und Tetrasaccharid-Sequenzen von N-Glycoproteinen mit β-D-mannosidischer Verknüpfung, *Liebigs Ann. Chem.*, 1047, 1983.

178. **Bundle, D. R. and Josephson, S.**, Artificial carbohydrate antigens: the synthesis of a tetrasaccharide hapten, a *Shigella flexneri* O-antigen repeating unit, *Carbohydr. Res.*, 80, 75, 1980.

179. **Horton, D. and Samneth, S.**, Synthesis of 8-(methoxycarbonyl)octyl glycosides of *O*-α-L-rhamnopyranosyl-(1→3)-*O*-α-L-rhamnopyranosyl-(1→3)-2-acetamido-2,6-dideoxy-D-glucopyranose; models for the antigen of *Pseudomonas aeruginosa* Fisher immunotype 5, *Carbohydr. Res.*, 103, C12, 1982.

180. **Matta, K. L., Rana, S. S., Piskorz, C. F., and Abbas, S. S.**, Synthesis of some oligosaccharides containing the *O*-α-L-fucopyranosyl-(1→2)-D-galactopyranosyl unit, *Carbohydr. Res.*, 131, 247, 1984.

181. **Nashed, M. A. and Anderson, L.,** Oligosaccharides from "standardized intermediates". Synthesis of a branched tetrasaccharide glycoside related to the blood group B determinant, *J. Am. Chem. Soc.,* 104, 7282, 1982.

182. **Paulsen, H., Kolář, Č., and Stenzel, W.,** Synthese der Trisaccharidkette der Determinante der Blutgruppensubstanz A, Typ 1, *Chem. Ber.,* 111, 2370, 1978.

183. **Bovin, N. V., Zurabyan, S. E., and Khorlin, A. Ya.,** Stereospecific glycosylation with 2-azido-2-deoxy-D-galactopyranose derivatives and synthesis of determinant oligosaccharide related to the blood-group A, type 1, *Izv. Akad. Nauk. SSSR, Ser. Khim.,* 1148, 1982.

184. **Paulsen, H., Stenzel, W., and Kolář, Č.,** Synthese von Oligosaccharid-Einheiten von Blutgruppensubstanzen mit Terminalen α-verknüpften D-Galactosamin-Gruppierungen, *Tetrahedron Lett.,* 2785, 1977.

185. **Bovin, N. V., Zurabyan, S. E., and Khorlin, A. Ya.,** Synthesis of blood-group A determinant oligosaccharides, *Bioorg. Khim.,* 7, 1271, 1981.

186. **Bovin, N. V., Zurabyan, S. E., and Khorlin, A. Ya.,** Stereoselectivity of glycosylation with derivatives of 2-azido-2-deoxy-D-galactopyranose. The synthesis of a determinant oligosaccharide related to blood-group A (Type 1), *Carbohydr. Res.,* 112, 23, 1983.

187. **Paulsen, H. and Kolář, Č.,** Synthese der Tetrasaccharidketten der Determinanten der Blutgruppensubstanzen A und B, *Angew. Chem.,* 90, 823, 1978.

188. **Derevitskaya, V. A., Novikova, O. S., Evstigneev, A. Yu., and Kochetkov, N. K.,** Synthesis of 2-acetamido-2-deoxy-3-*O*-[2-*O*-(α-L-fucopyranosyl)-β-D-galactopyranosyl]-D-glucose, a determinant oligosaccharide related to blood-group H, *Izv. Akad. Nauk SSSR Ser. Khim.,* 450, 1978.

189. **Paulsen, H. and Kolář, Č.,** Bausteine von Oligosacchariden. XVI. Synthese der Oligosaccharid Determinanten der Blutgruppensubstanzen der Type 1 des ABH-Systems, *Chem. Ber.,* 112, 3190, 1979.

190. **Paulsen, H. and Schnell, D.,** Synthese der Trisaccharid-Sequenz α-D-GlcpNAc-(1→4)-β-D-Gal-(1→4)-D-GlcpNAc aus blutgruppenaktiven Substanzen, *Chem. Ber.,* 114, 333, 1981.

191. **Wong, T. C. and Lemieux, R. U.,** The synthesis of derivatives of lactosamine and cellobiosamine to serve as probes in studies of the combining site of the monoclonal anti-I Ma antibody, *Can. J. Chem.,* 62, 1207, 1984.

192. **Jacquinet, J. C., Duchet, D., Milat, M. L., and Sinaÿ, P.,** Synthesis of blood-group substances. Part 11. Synthesis of the trisaccharide *O*-α-D-galactopyranosyl-(1→3)-*O*-β-D-galactopyranosyl-(1→4)-2-acetamido-2-deoxy-D-glucopyranose, *J. Chem. Soc. Perkin I,* 326, 1981.

193. **Garegg, P. J. and Oscarson, S.,** A synthesis of 8-methoxycarbonyloct-1-yl *O*-α-D-galactopyranosyl-(1→3)-*O*-β-D-galactopyranosyl-(1→4)-2-acetamido-2-deoxy-β-D-glucopyranoside, *Carbohydr. Res.,* 136, 207, 1985.

194. **Zollo, P. H., Jacquinet, J. C., and Sinaÿ, P.,** Chemical synthesis of the human blood-group P_1-antigenic determinant, *Carbohydr. Res.,* 122, 201, 1983.

195. **Nashed, M. A. and Anderson, L.,** Oligosaccharides from "standardized intermediates". Synthesis of a branched tetrasaccharide glycoside isomeric with the blood-group B, type 2 determinant, *Carbohydr. Res.,* 114, 43, 1983.

196. **Dahmén, J., Frejd, T., Magnusson, G., Noori, G., and Carlström, A.-S.,** Synthesis of spacer-arm, lipid, and ethyl glycosides of the terminal trisaccharide [α-D-Gal-(1→4)-β-D-Gal-(1→4)-β-D-GlcNAc] portion of the blood-group P_1 antigen: preparation of neoglycoproteins, *Carbohydr. Res.,* 129, 63, 1984.

197. **Jacquinet, J. C. and Sinaÿ, P.,** Synthese des substances de groupe sanguin IV. Synthese de 2-acetamide-2-desoxy-4-*O*-[2-*O*-(α-L-fucopyranosyl)-β-D-galactopyranosyl]-D-glucopyranose, porteur de la specificite H, *Tetrahedron,* 32, 1693, 1976.

198. **Paulsen, H. and Kolář, Č.,** Bausteine von Oligosacchariden. XX. Synthese der Oligosaccharid-Determinanten der Blutgruppensubstanzen der Type 2 des ABH-Systems. Diskussion der α-Glycosid-Synthese, *Chem. Ber.,* 114, 306, 1981.

199. **Rana, S. S., Vig, R., and Matta, K. L.,** Synthesis of *O*-α-L-fucopyranosyl-(1→2)-*O*-β-D-galactopyranosyl-(1→4)-2-acetamido-2-deoxy-D-glucopyranose. The H blood-group specific trisaccharide, *J. Carbohydr. Chem.,* 1, 261, 1982-1983.

200. **Milat, M.-L. and Sinaÿ, P.,** Synthesis of tetrasaccharide *O*-α-L-fucopyranosyl-(1→2)-[*O*-α-D-galactopyranosyl-(1→3)]-*O*-β-D-galactopyranosyl-(1→4)-2-acetamido-2-deoxy-D-glucopyranose, the antigenic determinant of human blood-group B (type 2), *Carbohydr. Res.,* 92, 183, 1981.

201. **Milat, M.-L. and Sinaÿ, P.,** Chemical synthesis of human blood group B antigenic determinant: type 2 tetrasaccharide, *Angew. Chem. Int. Ed. Engl.,* 18, 464, 1979.

202. **Paulsen, H. and Kolář, Č.,** Synthese der Tetrasaccharid-Kette der Type 2 der Determinanten der Blutgruppensubstanzen A und B, *Tetrahedron Lett.,* 2881, 1979.

203. **Jacquinet, J. C. and Sinaÿ, P.,** Synthesis of blood-group substances. 6. Synthesis of *O*-α-L-fucopyranosyl-(1→2)-*O*-β-D-galactopyranosyl-(1→4)-*O*-[α-L-fucopyranosyl-(1→3)]-2-acetamido-2-deoxy-α-D-glucopyranose, the postulated Lewis d antigenic determinant, *J. Org. Chem.,* 42, 720, 1977.

204. **Pougny, J.-R., Jacquinet, J. C., Nassr, M., Duchet, D., Milat, M. L., and Sinaÿ, P.,** A novel synthesis of 1,2-cis-disaccharides, *J. Am. Chem. Soc.,* 99, 6762, 1977.

205. **Paulsen, H. and Tietz, H.**, Synthese von Trisaccharide-Einheiten aus *N*-Acetylneuraminsäure und *N*-Acetyllactosamin, *Angew. Chem.*, 94, 934, 1982.

206. **Paulsen, H. and Tietz, H.**, Synthese eines Trisaccharides aus *N*-Acetylneuraminsäure und *N*-Acetyllactosamin, *Carbohydr. Res.*, 125, 47, 1984.

207. **Paulsen, H. and Tietz, H.**, Synthese eines *N*-Acetylneruaminsäure-haltigen Syntheseblocks. Kupplung zum *N*-Acetylneuraminsäuretetrasaccharid mit Trimethylsilyltriflat, *Carbohydr. Res.*, 144, 205, 1985.

208. **Paulsen, H. and Tietz, H.**, Herstellung eines *N*-Acetylneuraminsäure-haltigen Trisaccharides und dessen Verwendung in Oligosaccharidsynthesen, *Angew. Chem.*, 97, 118, 1985.

209. **Yoshikawa, M., Ikeda, Y., Takenaka, K., Torihara, M., and Kitagawa, I.**, Synthesis of ribostamycin. An application of a chemical conversion method from carbohydrate to aminocyclitol, *Chem. Lett.*, 2097, 1984.

210. **Oguri, S., Ishihara, H., and Tejima, S.**, A preparation of the branched trisaccharide 2-acetamido-2-deoxy-4-*O*-(β-D-galactopyranosyl)-3-*O*-(β-D-xylopyranosyl)-D-glucopyranose (3-*O*-β-D-xylopyranosyl-*N*-acetyllactosamine), *Chem. Pharm. Bull.*, 28, 35, 1980.

211. **Oguri, S., Ishihara, H., and Tejima, S.**, Synthesis of 2-acetamido-4-*O*-(2-acetamido-2-deoxy-β-D-glucopyranosyl)-2-deoxy-3-*O*-(α-L-fucopyranosyl)-D-glucopyranose (3-*O*-α-L-fucopyranosyl-di-*N*-acetylchitobiose), *Chem. Pharm. Bull.*, 28, 3196, 1980.

212. **Oguri, S. and Tejima, S.**, Synthesis of 2-acetamido-4-*O*-(2-acetamido-2-deoxy-β-D-glucopyranosyl)-2-deoxy-6-*O*-(α-L-fucopyranosyl)-D-glucopyranose (6-*O*-α-L-fucopyranosyl-di-*N*-acetylchitobiose), *Chem. Pharm. Bull.*, 29, 1629, 1981.

213. **Schwartz, D. A., Lee, H. H., Carver, J. P., and Krepinsky, J. J.**, Synthesis of model oligosaccharides of biological significance. 4. Synthesis of a fucosylated *N,N'*-diacetylchitobioside and related oligosaccharides, *Can. J. Chem.*, 63, 1073, 1985.

214. **Lee, H. H., Schwartz, D. A., Harris, J. F., Carver, J. P., and Krepinsky, J. J.**, Syntheses of model oligosaccharides of biological significance. 7. Synthesis of a fucosylated *N,N'*-diacetylchitobioside linked to bovine serum albumin and immunochemical characterization of rabbit antisera to this structure, *Can. J. Chem.*, 64, 1912, 1986.

215. **Derevitskaya, V. A., Novikova, O. S., and Kochetkov, N. K.**, Synthesis of 2-acetamido-2-deoxy-3,6-di-*O*-(β-D-galactopyranosyl)-D-glucose, *Izv. Akad. Nauk SSSR Ser. Khim.*, 1350, 1976.

216. **Lemieux, R. U. and Driguez, H.**, The chemical synthesis of 2-acetamido-2-deoxy-4-*O*-(α-L-fucopyranosyl)-3-*O*-(β-D-galactopyranosyl)-D-glucose. The Lewis a blood-group antigenic determinant, *J. Am. Chem. Soc.*, 97, 4063, 1975.

217. **Jacquinet, J. C. and Sinaÿ, P.**, Synthesis of blood-group substances. Part 8. A synthesis of the branched trisaccharide 2-acetamido-2-deoxy-4-*O*-(α-L-fucopyranosyl)-3-*O*-(β-D-galactopyranosyl)-D-glucopyranose, *J. Chem. Soc. Perkin I*, 319, 1979.

218. **Bovin, N. V., Zurabyan, S. E., and Khorlin, A. Ya.**, Chloroacetyl and 2-tetrahydrofuranyl groups as temporary protection in oligosaccharide synthesis, *Bioorg. Khim.*, 6, 242, 1980.

219. **Rana, S. S. and Matta, K. L.**, A facile synthesis of 2-acetamido-2-deoxy-4-*O*-α-L-fucopyranosyl-3-*O*-β-D-galactopyranosyl-D-glucopyranose, the Lewis a blood-group antigenic determinant, and related compounds, *Carbohydr. Res.*, 117, 101, 1983.

220. **Bovin, N. V., Ivanova, I. A., and Khorlin, A. Ya.**, Artificial carbohydrate antigens. Conjugation of the Le^a trisaccharide by the way of: oligosaccharide → glycosylated spacer → antigen, *Bioorg. Khim.*, 11, 662, 1985.

221. **Sykes, D. E., Rana, S. S., Barlow, J. J., and Matta, K. L.**, Modified assay-procedure for guanosine diphosphate-L-fucose: 2-acetamido-2-deoxy-β-D-glucopyranoside-(1→4)-α-L-fucosyltransferase with the aid of synthetic phenyl 2-acetamido-2-deoxy-4-*O*-α-L-fucopyranosyl-3-*O*-β-D-galactopyranosyl-β-D-glucopyranoside as a reference compound, *Carbohydr. Res.*, 112, 221, 1983.

222. **Lemieux, R. U., Bundle, D. R., and Baker, D. A.**, The properties of a "synthetic" antigen related to the human blood-group Lewis a, *J. Am. Chem. Soc.*, 97, 4076, 1975.

223. **Bovin, N. V., Zurabayan, S. E., and Khorlin, A. Ya.**, A simple synthesis of the determinant trisaccharide of the Le^a blood-group, *Bioorg. Khim.*, 6, 789, 1980.

224. **Jacquinet, J. C. and Sinaÿ, P.**, Synthesis of blood-group substances. Part 7. Synthesis of the branched trisaccharide *O*-α-L-fucopyranoyl-(1→3)-[*O*-β-D-galactopyranosyl-(1→4)]-2-acetamido-2-deoxy-D-glucopyranose, *J. Chem. Soc. Perkin I*, 314, 1979.

225. **Lönn, H.**, Synthesis of a tetra- and a nona-saccharide which contain α-L-fucopyranosyl groups and are part of the complex type of carbohydrate moiety of glycoproteins, *Carbohydr. Res.*, 139, 115, 1985.

226. **Kaifu, R. and Osawa, T.**, Synthesis of *O*-β-D-galactopyranosyl-(1→4)-*O*-(2-acetamido-2-deoxy-β-D-glucopyranosyl)-(1→2)-D-mannose and its interaction with various lectins, *Carbohydr. Res.*, 52, 179, 1976.

227. **Arnarp, J. and Lönngren, J.**, Synthesis of a tri-, a penta-, and a hepta-saccharide containing terminal *N*-acetyl-β-D-lactosaminyl residues, part of the "complex-type" carbohydrate moiety of glycoproteins, *J. Chem. Soc. Perkin I*, 2070, 1981.

228. **Arnarp, J., Baumann, H., Lönn, H., Lönngren, J., Nyman, H., and Ottosson, H.,** Synthesis of oligosaccharides that form parts of the complex type of carbohydrate moieties of glycoproteins. Three glycosides prepared for conjugation to proteins, *Acta Chem. Sand. B,* 37, 329, 1983.

229. **Sadozai, K. K., Kitajima, T., Nakahara, Y., Ogawa, T., and Kobata, A.,** Synthesis of a pentasaccharide hapten related to a monoantennary glycan chain of human chorionic gonadotropin, *Carbohydr. Res.,* 152, 173, 1986.

230. **Paulsen, H. and Lebuhn, R.,** Synthese von Pentasaccharid- und Octasaccharid-Sequenzen der Kohlen-hydrat-Kette von N-Glycoproteinen, *Carbohydr. Res.,* 125, 21, 1984.

231. **Ogawa, T., Kitajima, T., and Nukada, T.,** Synthesis of a nonahexosyl unit of a complex type of glycan chain of a glycoprotein, *Carbohydr. Res.,* 123, C8, 1983.

232. **Ogawa, T., Sugimoto, M., Kitajima, T., Sadozai, K. K., and Nukada, T.,** Total synthesis of a undecasaccharide: a typical carbohydrate sequence for the complex type of glycan chains of a glycoprotein, *Tetrahedron Lett.,* 5739, 1986.

233. **Lönn, H.,** Synthesis of a tri- and a hepta-saccharide which contain α-L-fucopyranosyl groups and are part of the complex type of carbohydrate moiety of glycoproteins, *Carbohydr. Res.,* 139, 105, 1985.

234. **Alais, J. and Veyeriéres, A.,** Synthesis of O-β-D-galactopyranosyl-(1→4)-O-2-acetamido-2-deoxy-β-D-glucopyranosyl-(1→3)-D-mannose, a postulated trisaccharide of human erythrocyte membrane sialoglyco-protein, *J. Chem. Soc. Perkin I,* 377, 1981.

235. **Sandozai, K. K., Nukada, T., Ito, Y., Nakahara, Y., Ogawa, T., and Kobata, A.,** Synthesis of a heptasaccharide hapten related to a biantennary glycan chain of human chorionic gonadotropin of a cho-riocarcinoma patient. A convergent approach, *Carbohydr. Res.,* 157, 101, 1986.

236. **Rana, S. S., Barlow, J. J., and Matta, K. L.,** Synthesis of p-nitrophenyl 2-acetamido-2-deoxy-4-O-β-D-galactopyranosyl-β-D-glucopyranoside, and p-nitrophenyl 6-O-(2-acetamido-2-deoxy-3-O- and -4-O-β-D-galactopyranosyl-β-D-glucopyranosyl)-α-D-mannopyranoside, *Carbohydr. Res.,* 113, 257, 1983.

237. **Alais, J. and Veyriéres, A.,** Blood-group Ii-active oligosaccharides. Synthesis of O-β-D-galactopyranosyl-(1→4)-O-(2-acetamido-2-deoxy-β-D-glucopyranosyl)-(1→6)-D-mannose, *Carbohydr. Res.,* 92, 310, 1981.

238. **Arnarp, J., Lönngren, J., and Ottosson, H.,** Synthesis of O-β-D-galactopyranosyl-(1→4)-O-(2-acetam-ido-2-deoxy-β-D-glucopyranosyl)-(1→6)-D-mannopyranose, *Carbohydr. Res.,* 98, 154, 1981.

239. **Ogawa, T. and Yamamoto, H.,** Synthesis of linear D-mannotetraose and D-mannohexaose, partial structures of the cell-surface D-mannan of *Candida albicans* and *Candida utilis, Carbohydr. Res.,* 104, 271, 1982.

240. **Lipták, A., Imre, J., and Nánási, P.,** Synthesis of O-α-D-mannopyranosyl-(1→2)-O-α-D-mannopyranosyl-(1→2)-D-mannose, the repeating unit of the 08-antigen of *Escherichia coli, Carbohydr. Res.,* 114, 35, 1983.

241. **Ogawa, T. and Nukada, T.,** Synthesis of a branched mannohexaoside, a part structure of a high-mannose-type glycan of a glycoprotein, *Carbohydr. Res.,* 136, 135, 1985.

242. **Ogawa, T. and Beppu, K.,** Synthesis of glycoglycerolipids: 3-O-Mannooligosyl-1,2-di-O-tetradecyl-*sn*-glycerol, *Agric. Biol. Chem.,* 46, 263, 1982.

243. **Srivastava, O. P. and Hindsgaul, O.,** Synthesis of the subterminally 6-O-phosphorylated trimannosides found on carbohydrate chains of lysosomal enzymes, *Can. J. Chem.,* 64, 2324, 1986.

244. **Takeda, T., Fujisawa, S., Ogihara, Y., and Hori, T.,** Synthesis of derivatives of the trisaccharide GlcNAcβ(1→2)Manα(1→3)Man, *Chem. Pharm. Bull.,* 33, 540, 1985.

245. **Ogawa, T. and Yamamoto, H.,** Synthesis of model linear mannohexaose for the backbone structure of fruit body polysaccharide of *Tremella fuciformis* and *Dictiophora indusiata* FISCH, *Agric. Biol. Chem.,* 49, 475, 1985.

246. **Tahir, S. H. and Hindsgaul, O.,** Substrates for the differentiation of the *N*-acetylglucosaminyltransferases. Synthesis of β-D-GlcNAc-(1→2)-α-D-Man-(1→6)-β-D-Man and β-D-GlcNAc-(1→2)-α-D-Man-(1→6)[α-D-Man-(1→3)]-β-D-Man glycosides, *Can. J. Chem.,* 64, 1771, 1986.

247. **Shah, R. N., Cumming, D. A., Grey, A. A., Carver, J. P., and Krepinsky, J. J.,** Synthesis of a tetrasaccharide of the extended core-region of the saccharide moiety of N-linked glycoproteins, *Carbohydr. Res.,* 153, 155, 1986.

248. **Lee, R. T. and Lee, Y. C.,** Synthesis of some O-D-xylosyl-D-mannoses and their derivatives, *Carbohydr. Res.,* 67, 389, 1978.

249. **Winnik, F. M., Carver, J. P., and Krepinsky, J. J.,** Synthesis of model oligosaccharides of biological significance. 2. Synthesis of a tetramannoside and of two lyxose-containing trisaccharides, *J. Org. Chem.,* 47, 2701, 1982.

250. **Arnarp, J., Haraldsson, M., and Lönngren, J.,** Synthesis of three oligosaccharides that form part of the complex type of carbohydrate moiety of glycoproteins, *Carbohydr. Res.,* 97, 307, 1981.

251. **Ogawa, T. and Nakabayashi, S.,** Synthesis of di- and trisaccharides having the sequences [β-D-GlcNAc]ₙ(1→X)-α-D-Man in the glycan of glycopeptides, *Agric. Biol. Chem.,* 45, 2329, 1981.

252. **Ogawa, T. and Sasajima, K.,** Synthesis of a model of an inner chain of cell-wall proteo-heteroglycan isolated from *Piricularia oryzae:* Branched D-manno-pentaosides, *Carbohydr. Res.,* 93, 67, 1981.

253. **Takeda, T., Kawarasaki, I., and Ogihara, Y.,** Studies on the structure of a polysaccharide from *Epiderphyton floccosum* and approach to a synthesis of the basic trisaccharide repeating units, *Carbohydr. Res.*, 89, 301, 1981.

254. **Eby, R. and Schuerch, C.,** The synthesis of antigenic determinants for yeast D-mannans and a linear (1→6)-α-D-gluco-D-mannan, and their protein conjugates, *Carbohydr. Res.*, 77, 61, 1979.

255. **Arnarp, J. and Lönngren, J.,** Synthesis of 3,6-di-*O*-(α-D-mannopyranosyl)-D-mannose, *Acta Chem. Scand.*, 32, 696, 1978.

256. **Ogawa, T., Katano, K., and Matsui, M.,** Regio- and stereo-controlled synthesis of core oligosaccharides of glycopeptides, *Carbohydr. Res.*, 64, C3, 1978.

257. **Ogawa, T., Katano, K., Sasajima, K., and Matsui, M.,** Synthetic studies on cell surface glycans. 3. Branching pentasaccharides of glycoprotein, *Tetrahedron*, 37, 2779, 1981.

258. **Ogawa, T. and Sasajima, K.,** Reconstruction of glycan chains of glycoprotein. Branching mannopentaoside and mannohexaoside, *Tetrahedron*, 37, 2787, 1981.

259. **Ogawa, T. and Sasajima, K.,** Synthesis of a branched D-mannopentaoside and a branched D-mannohexaoside: models of the outer chain of the glycan of soybean agglutinin, *Carbohydr. Res.*, 93, 53, 1981.

260. **Ogawa, T. and Sasajima, K.,** Synthesis of a branched D-mannopentaoside and a branched D-mannohexaoside: models of the inner core of cell-wall glycoproteins of *Saccharomyces cerevisiae*, *Carbohydr. Res.*, 93, 231, 1981.

261. **Winnik, F. M., Brisson, J. R., Carver, J. P., and Krepinsky, J. J.,** Synthesis of model oligosaccharides of biological significance. Synthesis of methyl 3,6-di-*O*-(α-D-mannopyranosyl)-α-D-mannopyranoside and the corresponding mannobiosides, *Carbohydr. Res.*, 103, 15, 1982.

262. **Winnik, F. M., Carver, J. P., and Krepinsky, J. J.,** Synthesis of model oligosaccharides of biological significance. 3. Synthesis of carbon-13 labelled trimannosides, *J. Labelled Compd. Radiopharm.*, 20, 983, 1983.

263. **Ogawa, T. and Nakabayashi, S.,** Synthesis of a hexasaccharide unit of a complex type of glycan chain of a glycoprotein, *Carbohydr. Res.*, 93, C1, 1981.

264. **Ogawa, T., Nakabayashi, S., and Kitajima, T.,** Synthesis of a hexasaccharide unit of a complex type of glycan chain of a glycoprotein, *Carbohydr. Res.*, 114, 225, 1983.

265. **Lönn, H. and Lönngren, J.,** Synthesis of a nona- and an undeca-saccharide that form part of the complex type of carbohydrate moiety of glycoproteins, *Carbohydr. Res.*, 120, 17, 1983.

266. **Arnarp, J., Haraldsson, M., and Lönngren, J.,** Synthesis of a nonasaccharide containing terminal *N*-acetyl-β-D-lactosaminyl residues, part of the "complex-type" carbohydrate moiety of glycoproteins, *J. Chem. Soc. Perkin I*, 1841, 1982.

267. **Sadozai, K. K., Ito, Y., Nukada, T., Ogawa, T., and Kobata, A.,** Synthesis of a heptasaccharide hapten related to an anomalous biantennary glycan-chain of human chorionic gonadotropin of a patient with choriocarcinoma. A stepwise approach, *Carbohydr. Res.*, 150, 91, 1986.

268. **Sadozai, K. K., Nukada, T., Ito, Y., Kobata, A., and Ogawa, T.,** Synthesis of a heptasaccharide hapten related to an anomalous biantennary glycan chain of human chorionic gonadotropin of a patient with choriocarcinoma, *Agric. Biol. Chem.*, 50, 251, 1986.

269. **Iversen, T. and Bundle, D. R.,** Antigenic determinants of *Salmonella* serogroups A and D$_1$. Synthesis of trisaccharide glycosides for use as artificial antigens, *Carbohydr. Res.*, 103, 29, 1982.

270. **Pinto, B. M. and Bundle, D. R.,** Antigenic determinant of *Salmonella* serogroup B. Synthesis of a trisaccharide glycoside for use as an artificial antigen, *Carbohydr. Res.*, 133, 333, 1984.

271. **Paulsen, H. and Lorentzen, J. P.,** Synthese von Tri- und Tetrasaccharide-Einheiten des O-Antigens aus *Aeromonas salmonicida*, *Tetrahedron Lett.*, 6043, 1985.

272. **Bundle, D. R. and Josephson, S.,** Artificial carbohydrate antigens. Synthesis and conformation of a *Shigella flexneri* trisaccharide hapten, *J. Chem. Soc. Perkin I*, 2736, 1979.

273. **Garegg, P. J., Hultberg, H., and Lindberg, C.,** Synthesis of *p*-trifluoroacetamidophenyl *O*-α-D-galactopyranosyl-(1→2)-*O*-α-D-mannopyranosyl-(1→4)-α-L-rhamnopyranoside, *Carbohydr. Res.*, 83, 157, 1980.

274. **Bock, K. and Meldal, M.,** Synthesis of the branchpoint tetrasaccharide of the O-specific determinant of *Salmonella* serogroup B, *Acta Chem. Scand. B*, 38, 71, 1984.

275. **Bock, K. and Meldal, M.,** Mercury iodide as a catalyst in oligosaccharide synthesis, *Acta Chem. Scand. B*, 37, 775, 1983.

276. **Garegg, P. J., Henrichson, C., Norberg, T., and Ossowski, P.,** Synthesis of trisaccharides related to *Salmonella* serogroup E O-antigenic polysaccharides, *Carbohydr. Res.*, 119, 95, 1983.

277. **Kochetkov, N. K., Torgov, V. I., Malysheva, N. N., and Shashkov, A. S.,** Synthesis of the pentasaccharide repeating unit of the O-specific polysaccharide from *Salmonella strasbourg*, *Tetrahedron*, 36, 1099, 1980.

278. **Backinowsky, L. V., Gomtsyan, A. R., Byramova, N. E., and Kochetkov, N. K.,** Synthesis of oligosaccharide fragments of *Shigella flexneri* O-specific polysaccharide. II. Synthesis of trisaccharide Glcα(1→3)Rhaα(1→2)Rhaα(1→0)Me and tetrasaccharide GlcNAcβ(1→2)(Glcα(1→3))-Rhaα(1→2)Rhaα(1→0)Me, *Bioorg. Khim.*, 11, 254, 1985.

279. **Gigg, J., Gigg, R., Payne, S., and Conant, R.,** The allyl group for protection in carbohydrate chemistry. 17. Synthesis of propyl *O*-(3,6-di-*O*-methyl-β-D-glucopyranosyl)-(1→4)-*O*-(2,3-di-*O*-methyl-α-L-rhamnopyranosyl)-(1→2)-3-*O*-methyl-α-L-rhamnopyranoside: the oligosaccharide portion of the major serologically active glycolipid from *Mycobacterium leprae, Chem. Phys. Lipids,* 38, 299, 1985.

280. **Fujiwara, T., Hunter, S. W., Cho, S.-N., Aspinall, G. O., and Brennan, P. J.,** Chemical synthesis and serology of disaccharides and trisaccharides of phenolic glycolipid antigens from the Leprosy Bacillus and preparation of disaccharide protein conjugate for serodiagnosis of leprosy, *Infect. Immun.,* 43, 245, 1984.

281. **Josephson, S. and Bundle, D.,** Artificial carbohydrate antigens: the synthesis of the tetrasaccharide repeating unit of *Shigella flexneri* O-antigen, *Can. J. Chem.,* 57, 3073, 1979.

282. **Wessel, H. P. and Bundle, D. R.,** Strategies for the synthesis of branched oligosaccharides of the *Shigella flexneri* 5a, 5b and variant X serogroups employing multifunctional rhamnose precursor, *J. Chem. Soc. Perkin I,* 2251, 1985.

283. **Jaworska, A. and Zamojski, A.,** A new method of oligosaccharide synthesis: rhamnotrioses, *Carbohydr. Res.,* 126, 205, 1984.

284. **Pozsgay, V., Nánási, P., and Neszmélyi, A.,** Synthesis, and carbon-13 N.M.R. study, of *O*-α-L-rhamnopyranosyl-(1→3)-*O*-α-L-rhamnopyranosyl-(1→2)-L-rhamnopyranose and *O*-α-L-rhamnopyranosyl-(1→3)-*O*-α-L-rhamnopyranosyl-(1→3)-L-rhamnopyranose, constituents of bacterial, cell-wall polysaccharides, *Carbohydr. Res.,* 90, 215, 1981.

285. **Iversen, T., Josephson, S., and Bundle, D. R.,** Synthesis of Streptococcal groups A, C and variant-A antigenic determinants, *J. Chem. Soc. Perkin I,* 2379, 1981.

286. **Lipták, A., Harangi, J., Batta, Gy., Seligmann, O., and Wagner, H.,** Synthesis and ^{13}C-N.M.R. spectroscopic investigation of three methyl rhamnotriosides, *Carbohydr. Res.,* 131, 39, 1984.

287. **Lipták, A., Szabó, L., Kerékgyártó, J., Harangi, J., Nánási, P., and Duddeck, H.,** Synthesis of the tetrasaccharide repeating-unit of the lipopolysaccharide isolated from *Pseudomonas maltophilia, Carbohydr. Res.,* 150, 187, 1986.

288. **Backinowksy, L. V., Gomtsyan, A. R., Byramova, N. E., Kochetkov, N. K., and Yankina, N. F.,** Synthesis of oligosaccharide fragments of *Shigella flexneri* O-specific polysaccharides. IV. The synthesis of the trisaccharide Glcα(1→3)Rhaα(1→3)Rhaα-OMe and tetrasaccharides Rhaα(1→2)(Glcα(1→3))Rhaα(1→3)Rhaα-OMe and GlcNAcβ(1→2) Rhaα(1→2)(Glcα(1→3))Rhaα-OMe. Localization of the O-factor V, *Bioorg. Khim.,* 11, 1562, 1985.

289. **Josephson, S. and Bundle, D. R.,** Artificial carbohydrate antigens: synthesis of rhamnose trisaccharide and disaccharide haptens common to *Shigella flexneri* O-antigens, *J. Chem. Soc. Perkin I.,* 297, 1980.

290. **Tsvetkov, Yu. E., Byramova, N. E., Backinowsky, L. V., Kochetkov, N. K., and Yankina, N. F.,** Synthesis of a repeating unit of the basic chain of *Shigella flexneri* O-antigenic polysaccharides, *Bioorg. Khim.,* 12, 1213, 1986.

291. **Bayramova, N. E., Tsvetkov, Y. E., Backinowsky, L. V., and Kochetkov, N. K.,** Synthesis of the basic chain of the O-specific polysaccharides of *Shigella flexneri, Carbohydr. Res.,* 137, C8, 1985.

292. **Kochetkov, N. K., Byramova, N. E., Tsvetkov, Yu. E., and Backinowsky, L. V.,** Synthesis of the O-specific polysaccharide of *Shigella flexneri, Tetrahedron,* 41, 3363, 1985.

293. **Byramova, N. E., Tsvetkov, Yu. E., Backinowsky, L. V., and Kochetkov, N. K.,** Synthesis of the O-specific polysaccharide of *Shigella flexneri.* 5. Synthesis of *O*-(4,6-di-*O*-benzoyl-2-deoxy-2-phthaloimido-β-D-glucopyranosyl)-(1→2)-*O*-(3,4-di-*O*-benzoyl-α-L-rhamnopyranosyl)-(1→2)-*O*-(3,4-di-*O*-benzoyl-α-L-rhamnopyranosyl)-(1→3)-4-*O*-benzoyl-1,2-*O*-[1-(exo-cyano)ethylidene]-β-L-rhamnopyranoside-suitable monomer for polycondensation, *Izv. Akad. Nauk SSSR Ser. Khim.,* 1145, 1985.

294. **Paulsen, H. and Kutschker, W.,** Synthese von β-rhamnosidish verknüpften Oligosacchariden des Lipopolysaccharides aus *Shigella flexneri* Serotyp 6, *Carbohydr. Res.,* 120, 25, 1983.

295. **Paulsen, H. and Kutschker, W.,** Bausteine von Oligosacchariden, XLV. Synthese einer verzweigten Tetrasaccharid-Einheit der O-spezifischen Kette des Lipopolysaccharides aus *Shigella flexneri* Serotyp 6, *Liebigs Ann. Chem.,* 557, 1983.

296. **Paulsen, H. and Lockhoff, O.,** Synthese β-D-Mannopyranosyl- und 2-Azido-2-desoxy-α-D-glucopyranosyl-haltiger Disaccharid-Halogenid-Bausteine, *Tetrahedron Lett.,* 4027, 1978.

297. **Paulsen, H. and Lockhoff, O.,** Bausteine von Oligosacchariden, XXIX. Synthese des Trisaccharids aus *N*-Acetylglucosamin, Galactose und Rhamnose einer O-determinanten Kette von *Escherichia coli.* Abhängigkeit der Stereoselektivität der α-Glycosidsynthese von der Reaktivität des Pyranosylhalogenids, *Chem. Ber.,* 114, 3079, 1981.

298. **Paulsen, H. and Lockhoff, O.,** Bausteine von Oligosacchariden, XXX. Neue effektive β-Glycosidsynthese für Mannose-Glycoside Synthesen von Mannose-haltigen Oligosacchariden, *Chem. Ber.,* 114, 3102, 1981.

299. **Schwarzenbach, D. and Jeanloz, R. W.,** Synthesis of part of the antigenic repeating-unit of *Streptococcus pneumoniae* type II, *Carbohydr. Res.,* 90, 193, 1981.

300. **Schwarzenbach, D. and Jeanloz, R. W.,** The synthesis of part of the repeating unit of a pneumococcal polysaccharide, *Carbohydr. Res.,* 77, C5, 1979.

301. **Backinowsky, L. V., Gomtsyan, A. R., Byramova, N. E., and Kochetkov, N. K.**, Synthesis of oligosaccharide fragments of *Shigella flexneri* O-specific polysaccharides. The synthesis of two branched trisaccharides, *Bioorg. Khim.*, 10, 79, 1984.

302. **Paulsen, H. and Lorentzen, J. P.**, Synthese von verzweigten Tri- und Tetrasaccharidsequenzen der "Repeating Unit" der O-specifischen Kette des Lipopolysaccharides aus *Aeromonas salmonicida, Liebigs Ann. Chem.*, 1586, 1986.

303. **Fügedi, P., Lipták, A., Nánási, P., and Neszmélyi, A.**, Synthesis of 4-*O*-α-D-galactopyranosyl-L-rhamnose and 4-*O*-α-D-galactopyranosyl-2-*O*-β-D-glucopyranosyl-L-rhamnose using dioxolane-type benzylidene acetals as temporary protecting groups, *Carbohydr. Res.*, 80, 233, 1980.

304. **Collins, P. M. and Munasinghe, V. R. N.**, The photochemistry of carbohydrate derivatives. Part 7. The synthesis of methyl 3,4-di-*O*-(β-D-glucopyranosyl)-α-L-rhamnopyranoside from photolabile methyl 2,3-*O*-(2-nitrobenzylidene)-α-L-rhamnopyanoside, *J. Chem. Soc. Perkin I*, 1879, 1983.

305. **Collins, P. M. and Munasinghe, V. R. N.**, Synthesis of some branched trisaccharides using photolabile *O*-nitrobenzylidene acetals as temporary protecting groups, *J. Chem. Soc. Chem. Commun.*, 362, 1981.

306. **Baines, F. C., Collins, P. M., and Farnia, F.**, Synthesis of methyl 3-*O*-(α- and -β-D-galactopyranosyl)-4-*O*-(β-D-glucopyranosyluronic acid)-α-L-rhamnopyranoside using photolabile methyl 2,3-*O*-(2-nitrobenzylidene)-α-L-rhamnopyranoside, *Carbohydr. Res.*, 136, 27, 1985.

307. **Byramova, N. E., Backinowsky, L. V., and Kochetkov, N. K.**, Synthesis of the basic chain of O-antigenic polysaccharides of *Shigella flexneri*. 3. Synthesis of 4-*O*-benzoyl-3-*O*-(2-*O*-acetyl-3,4-di-*O*-benzoyl-α-L-rhamnopyranosyl)-1,2-*O*-[1-(exo-cyano)ethylidene]-β-L-rhamnopyranose, *Izv. Akad. Nauk SSSR Ser. Khim.*, 1134, 1985.

308. **Takeo, K., Nakaji, T., and Shinmitsu, K.**, Synthesis of lycotetraose, *Carbohydr. Res.*, 133, 275, 1984.

309. **Lichtenthaler, F. W. and Kaji, E.**, A facile access to trisaccharides with central β-D-mannose, α-D-glucosamine, and β-D-mannosamine units, *Liebigs Ann. Chem.*, 1659, 1985.

310. **Augé, C. and Veyriéres, A.**, Studies in oligosaccharide chemistry. Part 8. Synthesis of lacto-N-triose I, a core chain trisaccharide of human blood-group substances, *J. Chem. Soc. Perkin I.*, 1343, 1977.

311. **Lemieux, R. U., Abbas, S. Z., and Chung, B. Y.**, Synthesis of core chain trisaccharides related to human blood group antigenic determinant, *Can. J. Chem.*, 60, 68, 1982.

312. **Kohata, K., Abbas, S. A., and Matta, K. L.**, Synthetic mucin fragments: methyl 3-*O*-(2-acetamido-2-deoxy-β-D-glucopyranosyl)-β-D-galactopyranoside and methyl 3-*O*-(2-acetamido-2-deoxy-3-*O*-β-D-galactopyranosyl-β-D-glucopyranosyl)-β-D-galactopyranoside, *Carbohydr. Res.*, 132, 127, 1984.

313. **Augé, C., David, S., and Veyriéres, A.**, Molecular basis of the human I, blood-group systems. Contribution to the problem from the synthesis of I-active oligosaccharides, *Nouv. J. Chim.*, 3, 491, 1979.

314. **Dahmén, J., Gnosspelius, G., Larsson, A. C., Lave, T., Noori, G., Palsson, K., Frejd, T., and Magnusson, G.**, Synthesis of di-, tri-, and tetra-saccharides corresponding to receptor structures recognised by *Streptococcus pneumoniae, Carbohydr. Res.*, 138, 17, 1985.

315. **Matta, K. L. and Barlow, J. J.**, Synthesis of *p*-nitrophenyl 6-*O*-(2-acetamido-2-deoxy-β-D-glucopyranosyl)-β-D-galactopyranoside and p-nitrophenyl *O*-β-D-galactopyranosyl-(1→3)-*O*-(2-acetamido-2-deoxy-β-D-glucopyranosyl)-(1→6)-β-D-galactopyranoside, *Carbohydr. Res.*, 53, 209, 1977.

316. **Torgov, V. I., Panosyan, C. A., and Shibaev, V. N.**, Synthesis of moraprenyl pyrophosphate trisaccharide, a precursor in the biosynthesis of main chain of *Salmonella* serogroup C₂ and C₃ O-specific polysaccharides, and its isomer, *Bioorg. Khim.*, 12, 559, 1986.

317. **Kochetkov, N. K., Dmitriev, B. A., Nikolaev, A. V., and Byramova, N. E.**, Synthesis of antigenic bacterial polysaccharides and their fragments. 5, *Izv. Akad. Nauk SSSR Ser. Khim.*, 1609, 1977.

318. **Betaneli, V. I., Ovchinnikov, M. V., Backinowsky, L. V., and Kochetkov, N. K.**, Practical synthesis of *O*-β-D-mannopyranosyl-, *O*-α-D-mannopyranosyl-, and *O*-β-D-glucopyranosyl-(1→4)-α-L-rhamnopyranosyl-(1→3)-D-galactoses, *Carbohydr. Res.*, 84, 211, 1980.

319. **Kochetkov, N. K., Dmitriev, B. A., and Chernyak, A. Ya.**, Synthesis of antigenic bacterial polysaccharides and their fragments. VII. Synthesis of *p*-aminophenyl-3-*O*-[4-*O*-(β-D-glucopyranosyl)-α-L-rhamnopyranosyl]-β-D-galactopyranoside and its coupling to protein and sepharose, *Bioorg. Khim.*, 3, 752, 1977.

320. **Kochetkov, N. K., Dmitriev, B. A., Chernyak, A. Ya., and Byramova, N. E.**, Synthesis of antigenic bacterial polysaccharides and their fragments. 2. Synthesis of D-glucopyranosyl-(1→4)-α-L-rhamnopyranosyl-(1→3)-D-galactose, *Izv. Akad. Nauk SSSR Ser. Khim.*, 2331, 1974.

321. **Torgov, V. I. and Chernyak, A. Ya.**, Synthesis of isomeric glucopyranosyl-rhamnopyranosyl-galactopyranosyl derivatives, *Izv. Akad. Nauk SSSR Ser. Khim.*, 445, 1975.

322. **Kochetkov, N. K., Klimov, E. M., and Torgov, V. I.**, Synthesis of antigenic bacterial polysaccharides and their fragments. 3. Synthesis of α-D-mannopyranosyl-(1→4)-α-L-rhamnopyranosyl-(1→3)-D-galactopyranose, *Izv. Akad. Nauk SSSR Ser. Khim.*, 165, 1976.

323. **Torgov, V. I., Shibaev, V. N., Shashkov, A. S., and Kochetkov, N. K.**, Synthesis of bacterial antigens and their fragments. 12. Synthesis and ¹³C NMR spectra of di- and trisaccharide fragments of O-antigens from *Salmonella* serogroups A, B and D₁, *Bioorg. Khim.*, 6, 1860, 1980.

324. **Garegg, P. J. and Norberg, T.**, Synthesis of the biological repeating units of *Salmonella* serogroups A, B, and D₁ O-antigenic polysaccharides, *J. Chem. Soc. Perkin 1*, 2973, 1982.

325. **Kochetkov, N. K., Dmitriev, B. A., Malysheva, N. N., Chernyak, A. Ya., Klimov, E. M., Bayramova, N. E., and Torgov, V. I.**, Synthesis of *O*-β-D-mannopyranosyl-(1→4)-*O*-α-L-rhamnopyranosyl-(1→3)-D-galactopyranose, the trisaccharide repeating-unit of the O-specific polysaccharide from *Salmonella anatum*, *Carbohydr. Res.*, 45, 283, 1975.

326. **Kochetkov, N. K., Dmitriev, B. A., and Nikolaev, A. V.**, Synthesis of antigenic bacterial polysaccharides and their fragments. 6. Synthesis of β-D-mannopyranosyl-(1→4)-α-L-rhamnopyranosyl-(1→3)-D-galactose, the repeating-unit of O-specific polysaccharide from *Salmonella newington*, *Izv. Akad. Nauk SSSR Ser. Khim.*, 2578, 1977.

327. **Torgov, V. I., Chekunchikov, V. N., Shibaev, V. N., and Kochetkov, N. K.**, Synthesis of bacterial antigenic polysaccharide and their fragments. 13. Synthesis of trisaccharide repeating unit of O-antigenic polysaccharides from *Salmonella anatum* and *Salmonella newington* with radioactive label at galactose residue, *Bioorg. Khim.*, 7, 401, 1981.

328. **Kochetkov, N. K., Dmitriev, B. A., Chernyak, A. Ya., and Levinsky, A. B.**, A new type of carbohydrate-containing synthetic antigen: synthesis of carbohydrate-containing polysaccharide copolymers with the specificity of 0:3 and 0:4 factors of *Salmonella*, *Carbohydr. Res.*, 110, C16, 1982.

329. **Chernyak, A. Ya., Levinsky, A. B., Dmitriev, B. A., and Kochetkov, N. K.**, A new type of carbohydrate-containing synthetic antigen: synthesis of carbohydrate-containing polyacrylamide copolymers having the specificity of 0:3 and 0:4 factors of *Salmonella*, *Carbohydr. Res.*, 128, 269, 1984.

330. **Kochetkov, N. K., Dmitriev, B. A., Chernyak, A. Ya., Pokrovskii, V. I., and Tendetnik, Yu. Ya.**, New type of carbohydrate-containing synthetic antigens. Synthesis and immunochemical properties of a carbohydrate-containing copolimer with the specificity of factor 0:3 of *Salmonella* bacteria of serological group E, *Dokl. Akad. Nauk. SSSR*, 263, 1277, 1982.

331. **Kochetkov, N. K., Dmitriev, B. A., Chernyak, A. Ya., Pokrovsky, V. I., and Tendetnik, Yu. Ya.**, Synthesis of antigenic bacterial polysaccharides and their fragments. 11. Synthesis of an artificial antigen comprising biological repeating unit of the polysaccharide from *Salmonella newington* as haptenic groups, *Bioorg. Khim.*, 5, 217, 1979.

332. **Nikolaev, A. V., Shashkov, A. S., Dmitriev, B. A. and Kochetkov, N. K.**, Synthesis of antigenic bacterial polysaccharides and their fragments. 14. Formation of trisaccharide with β-L-rhamnopyranosidic linkage upon glycosylation of 4,6-*O*-ethylidene-1,2-*O*-isopropylidene-α-D-galactopyranose by hexa-*O*-acetyl-β-D-mannopyranosyl-(1→4)-α-L-rhamnopyranosyi bromide, *Bioorg. Khim.*, 7, 914, 1981.

333. **Kochetkov, N. K., Dmitriev, B. A., Chizhov, O. S., Klimov, E. M., Malysheva, N. N., Chernyak, A. Ya., Bayramova, N. E., and Torgov, V. I.**, Synthesis of derivatives of the trisaccharide repeating unit of the O-specific polysaccharide from *Salmonella anatum*, *Carbohydr. Res.*, 33, C5, 1974.

334. **Kochetkov, N. K., Dmitriev, B. A., Chizhov, O. S., Klimov, E. M., Malysheva, N. N., Torgov, B. I., Chernyak, A. Ya., and Byramova, N. E.**, Synthesis of antigenic bacterial polysaccharides and their fragments. 1. Synthesis of the repeating unit of the polysaccharide from *Salmonella anatum*, *Izv. Akad. Nauk SSSR Ser. Khim.*, 1386, 1974.

335. **Shibaev, V. N., Csekuncsikov, V. N., and Kochetkov, N. K.**, Synthesis of antigenic bacterial polysaccharides and their fragments. 9, *Izv. Akad. Nauk SSSR Ser. Khim.*, 2124, 1978.

336. **Lipták, A., Nánási, P., Neszmélyi, A., Riess-Maurer, I., and Wagner, H.**, Carbohydrate components of flavonol triosides: a convenient synthesis of *O*-α-L-rhamnopyranosyl-(1→3)-*O*-α-L-rhamnopyranosyl-(1→6)-D-galactose and *O*-α-L-rhamnopyranosyl-(1→2)-*O*-α-L-rhamnopyranosyl-(1→6)-D-galactose, *Carbohydr. Res.*, 93, 43, 1981.

337. **Lipták, A. and Nánási, P.**, A convenient synthesis of *O*-α-L-rhamnopyranosyl-(1→4)-*O*-α-L-rhamnopyranosyl-(1→6)-D-galactopyranose nona-acetate, *Carbohydr. Res.*, 44, 313, 1975.

338. **Wagner, H., Lipták, A., and Nánási, P.**, *O*-α-L-Rhamnopyranosyl-(1→4)-*O*-α-L-rhamnopyranosyl-(1→6)-D-galactopyranose nonacetate. Synthesis of the carbohydrate component of rhamnazin-3-*O*-trioside, *Acta Chim. Acad. Sci. Hung.*, 89, 405, 1976.

339. **Gent, P. A. and Gigg, R.**, Synthesis of 3-*O*-{6-*O*-[6-*O*-(α-D-galactopyranosyl)-α-D-galactopyranosyl]-β-D-galactopyranosyl}-1,2-di-*O*-stearoyl-L-glycerol, a "Trigalactosyl diglyceride", *J. Chem. Soc. Perkin 1*, 1779, 1975.

340. **Eby, R. and Schuerch, C.**, Synthesis and characterization of methyl 2-*O*-β-D-galactopyranosyl-β-D-galactopyranoside and methyl 2-*O*-(2-*O*-β-D-galactopyranosyl-β-D-galactopyranosyl)-β-D-galactopyranoside, *Carbohydr. Res.*, 92, 149, 1981.

341. **Kováč, P., Taylor, R. B., and Glaudemans, C. P. J.**, General synthesis of (1→3)-β-D-galacto oligosaccharides and their methyl β-glycosides by a stepwise or a blockwise approach, *J. Org. Chem.*, 50, 5323, 1985.

342. **Chowdhary, M. S., Navia, J. L., and Anderson, L.**, The assembly of oligosaccharides from "standardized intermediates": β-(1→3)-linked oligomers of D-glactose, *Carbohydr. Res.*, 150, 173, 1986.

343. El Shenawy, H. A. and Schuerch, C., Synthesis and characterization of propyl *O*-β-D-galactopyranosyl-(1→4)-*O*-β-D-galactopyranosyl-(1→4)-α-D-galactopyranoside, *Carbohydr. Res.*, 131, 239, 1984.

344. Koeners, H. J., Verhoeven, J., and van Boom, J. H., Application of levulinic acid ester as a protective groups in the synthesis of oligosaccharides, *Recl. Trav. Chim. Pays-Bas*, 100, 65, 1981.

345. Kováč, P., Yeh, H. J. C., and Glaudemans, C. P. J., Synthesis of methyl *O*-(3-deoxy-3-fluoro-β-D-galactopyranosyl)-(1→6)-β-D-galactopyranoside and methyl *O*-(3-deoxy-3-fluoro-β-D-galactopyranosyl)-(1→6)-*O*-β-D-galactopyranosyl-(1→6)-β-D-galactopyranoside, *Carbohydr. Res.*, 140, 277, 1985.

346. Nashed, E. M. and Glaudemans, C. P. J., Synthesis of 2,3-epoxypropyl β-glycosides of (1→6)-β-D-galactopyranooligosaccharides and their binding to monoclonal anti-galactan IgA J539 and IgA X24, *Carbohydr. Res.*, 158, 125, 1986.

347. Koeners, H. J., Verhoeven, J., and Van Boom, J. H., Synthesis of oligosaccharides by using levulinic ester as an hydroxyl protecting group, *Tetrahedron Lett.*, 21, 381, 1980.

348. Bhattacharjee, A. K., Zissis, E., and Glaudemans, C. P. J., Synthesis of 6-*O*-(6-*O*-β-D-galactopyranosyl-β-D-galactopyranosyl)-D-galactopyranose by use of 2,3,4-tri-*O*-acetyl-6-*O*-(chloroacetyl)-α-D-galactopyranosyl bromide, a key intermediate for the solid-phase synthesis of β-(1→6)-linked D-galactopyranans, *Carbohydr. Res.*, 89, 249, 1981.

349. Srivastava, V. K., Sondheimer, S. J., and Schuerch, C., Synthesis and characterization of methyl 6-*O*-β-D-galactopyranosyl-β-D-galactopyranoside and methyl *O*-β-D-galactopyranosyl-(1→6)-*O*-β-D-galactopyranosyl-(1→6)-β-D-galactopyranoside, *Carbohydr. Res.*, 86, 203, 1980.

350 Kováč, P., Efficient chemical synthesis of methyl β-glycosides of β-(1→6)-linked D-galacto-oligosaccharides by a stepwise and a blockwise approach, *Carbohydr. Res.*, 153, 237, 1986.

351. Kováč, P., Systematic chemical synthesis of (1→6)-β-D-galacto-oligosaccharides and related compounds, *Carbohydr. Res.*, 144, C12, 1985.

352. Kováč, P. and Glaudemans, C. P. J., Synthesis of specifically fluorinated methyl β-glycosides of (1→6)-β-D-galactooligosaccharides, *Carbohydr. Res.*, 123, C29, 1983.

353. Paulsen, H. and Bünsch, A., Bausteine von Oligosacchariden. XXXVI. Reaktivitätuntersuchungen bei Tri-und Pentasaccharidsynthesen. Verbesserte Synthese der Pentasaccharidkette des Forssman-Antigens, *Liebigs Ann. Chem.*, 2204, 1981.

354. Paulsen, H., Kolář, Č., and Stenzel, W., Bausteine von Oligosacchariden. XI. Synthese α-glycosidisch verknüpfter Disaccharide der 2-Amino-2-desoxy-D-galactopyranose, *Chem. Ber.*, 111, 2358, 1978.

355. Paulsen, H. and Bünsch, A., Synthese der Pentasaccharidkette des Forssman-Antigens, *Angew. Chem.*, 92, 929, 1980.

356. Aspinall, G. O., Chatterjee, D., and Khondo, L., The hex-5-enose degradation: zinc dust cleavage of 6-deoxy-6-iodo-α-D-galactopyranosidic linkages in methylated di- and trisaccharides, *Can. J. Chem.*, 62, 2728, 1984.

357. Temeriusz, A., Radomski, J., Stepinski, J., and Piekarska, B., The Koenigs-Knorr condensation of methyl 4,6-*O*-benzylidene-β-D-galactopyranoside with 2,3,4,6-tetra-*O*-acetyl-α-D-galactopyranosyl bromide, *Carbohydr. Res.*, 142, 146, 1985.

358. Koeners, H. J., Verdegaal, C. H. M., and van Boom, J. H., 4,4-(Ethylenedithio)pentanoyl: a masked levulinoyl protective group in the synthesis of oligosaccharides, *Recl. Trav. Chim. Pays-Bas*, 100, 118, 1981.

359. Augé, C. and Veyriéres, A., Stannylene derivatives in glycoside synthesis. Application to the synthesis of the blood-group B antigenic determinant, *J. Chem. Soc. Perkin I*, 1825, 1979.

360. Takeo, K., Fukatsu, T., and Okushio, K., Synthesis of solatriose, *Carbohydr. Res.*, 121, 328, 1983.

361. Augé, C. and Veyriéres, A., Synthesis of 3-*O*-(2-acetamido-2-deoxy-β-D-glucopyranosyl)-α-D-galactopyranose ("Lacto-N-biose II") and 3,4-di-*O*-(2-acetamido-2-deoxy-β-D-glucopyranosyl)-D-galactopyranose, *Carbohydr. Res.*, 54, 45, 1977.

362. David, S. and Veyriéres, A., The synthesis of 3,6-di-*O*-(2-acetamido-2-deoxy-β-D-glucopyranosyl)-D-galactose, a branched trisaccharide reported as a hydrolysis product of blood-group substances, *Carbohydr. Res.*, 40, 23, 1975.

363. Paulsen, H. and Lockhoff, O., Bausteine von Oligosacchariden. XXXI. Synthese der Repeating-Unit der O-spezifischen Kette des Lipopolysaccharides des Bakteriums *Escherichia coli* 075, *Chem. Ber.*, 114, 3115, 1981.

364. Riess-Maurer, I., Wagner, H., and Lipták, A., Synthesis of kaempferol-3-*O*-(3″,4″-di-*O*-α-L-rhamnopyranosyl)-β-D-galactopyranoside and its comparison with natural ascaside isolated from *Astragalus caucasicus*, *Z. Naturforsch.*, 36b, 257, 1981.

365. Lipták, A. and Nánási, P., A general method for the preparation of 2,6-di-*O*-glycosyl-hexoparanoses. Synthesis of *O*-β-D-galactopyranosyl-(1→2)-*O*-α-L-rhamnopyranosyl-(1→6)-D-galactopyranose decaacetate, *Tetrahedron Lett.*, 921, 1977.

366. Collins, P. M. and Eder, H., The photochemistry of carbohydrate derivatives. Part 6. Synthesis of 3-(methoxycarbonyl)propyl pyranosides of 2,3-di-*O*-(β-D-galactopyranosyl)-α-D-galactose and 3-*O*-(α-D-galactopyranosyl)-2-*O*-(β-D-galactopyranosyl)-α-D-galactose using photolabile 0-(2-nitrobenzylidene) acetals as temporary blocking groups, *J. Chem. Soc. Perkin I*, 927, 1983.

367. **Lemieux, R. U. and Driguez, H.**, The chemical synthesis of 2-*O*-(α-L-fucopyranosyl)-3-*O*-(α-D-galactopyranosyl)-D-galactose. The terminal structure of the blood-group B antigenic determinant, *J. Am. Chem. Soc.*, 97, 4069, 1975.

368. **Jaquinet, J. C. and Sinaÿ, P.**, Synthese des substances de groupe sanguin-IX. Une synthese du 2-O-(α-L-fucopyranosyl)-3-*O*-(α-D-galactopyranosyl)-α-D-galactose, le determinant antigenique du groupe sanguin B, *Tetrahedron*, 35, 365, 1979.

369. **David, S., Lubineau, A., and Vatéle, J.-M.**, The synthesis by the cycloaddition method of the trisaccharide antigenic determinant of the A human blood group system, and six related trisaccharides, one of them active in the "Acquired B" system, *Nouv. J. Chim.*, 4, 547, 1980.

370. **Subero, C., Jimeno, Ma. L., Alemany, A., and Martin-Lomas, M.**, A new synthesis of 2-*O*-α-L-fucopyranosyl-3-*O*-α-D-galactopyranosyl-D-galactose, *Carbohydr. Res.*, 126, 326, 1984.

371. **David, S., Lubineau, A., and Vatéle, J. M.**, Chemical synthesis of 2-*O*-(α-L-fucopyranosyl)-3-*O*-(2-acetamido-2-deoxy-α-D-galactopyranosyl)-D-galactose, the terminal structure in the blood-group A antigenic determinant, *J. Chem. Soc. Chem. Commun.*, 535, 1978.

372. **Paulsen, H. and Paal, M.**, Lewissäure-Katalysierte Synthesen von Di- und Trisaccharid-Sequenzen der *O*-Glycoproteine. Anwendung von Trimethylsylyltrifluoromethanesulfonate, *Charbohydr. Res.*, 135, 53, 1984.

373. **Grundler, G. and Schmidt, R. R.**, Anwendung des Trichloroacetamidätverfahrens auf 2-Desoxy-2-phthalimido-D-glucose-Derivate. Synthese von Oligosacchariden der "Core-Region" von *O*-Glycoproteinen des Mucin-Typs, *Carbohydr. Res.*, 135, 203, 1985.

374. **Lemieux, P. U. and Burzynska, M. H.**, Synthesis of β DGlcNAc(1→6) and β DGal(1→4) β DGlcNAc(1→6) derivatives of the T$_N$ (α DGalNAc) human blood group determinant, *Can. J. Chem.*, 60, 76, 1982.

375. **Rana, S. S., Barlow, J. J., and Matta, K. L.**, Synthesis of phenyl *O*-α-L-fucopyranosyl-(1→2)-*O*-β-D-galactopyranosyl-(1→3)-2-acetamido-2-deoxy-α-D-galactopyranoside, *Carbohydr. Res.*, 87, 99, 1980.

376. **Abbas, S. A. and Matta, K. L.**, Synthetic mucin fragments: benzyl 2-acetamido-3-*O*-[3-*O*-(2-acetamido-2-deoxy-β-D-glucopyranosyl)-β-D-galactopyranosyl]-2-deoxy-α-D-galactopyranoside, *Carbohydr. Res.*, 132, 137, 1984.

377. **Paulsen, H., Hasenkamp, T., and Paal, M.**, Entwicklung eines Syntheseblocks der 3-*O*-β-D-Galactopyranosyl-D-galactopyranose, *Carbohydr. Res.*, 144, 45, 1985.

378. **Catelani, G., Marra, A., Paquet, F., and Sinaÿ, P.**, Chemical synthesis of the desialylated human Cad-antigenic determinant, *Carbohydr. Res.*, 155, 131, 1986.

379. **Catelani, G., Marra, A., Paquet, F., and Sinaÿ, P.**, Synthesis of trisaccharide derivatives with the sequence of the desialylated Cad blood group determinant, *Gaz. Chim. Ital.*, 115, 565, 1985.

380. **Nashed, M. A. and Anderson, L.**, Oligosaccharides from "standardized intermediates". The 2-amino-2-deoxy-D-galactose analog of the blood-group O(H) determinant, type 2, and its precursors, *Carbohydr. Res.*, 114, 53, 1983.

381. **Horito, S., Lorentzen, J. P., and Paulsen, H.**, Bausteine von Oligosacchariden, LXXVII. Synthese einer Trisaccharidenheit des Kapselpolysaccharides von *Streptococcus pneumoniae* Typ 4, *Liebigs Ann. Chem.*, 1880, 1986.

382. **Paulsen, H. and Bünsch, H.**, Bausteine von Oligosacchariden, XXXII. Synthese der verzweigten Pentasaccharid-Einheit der O-spezifischen Seitenkette des Lipopolysaccharides von *Shigella dysenteriae*, *Chem. Ber.*, 114, 3126, 1981.

383. **Paulsen, H. and Bünsch, H.**, Synthese der Pentasaccharid-Sequenz der Repeating-Unit der O-spezifischen Seitenkette des Lipopolysaccharides von *Shigella dysenteriae*, *Tetrahedron Lett.*, 22, 47, 1981.

384. **Abbas, S. A., Barlow, J. J., and Matta, K. L.**, Synthesis of benzyl 2-acetamido-3,6-di-*O*-(2-acetamido-2-deoxy-β-D-glucopyranosyl)-2-deoxy-α-D-galactopyranoside, *Carbohydr. Res.*, 113, 63, 1983.

385. **Piskorz, C. F., Abbas, S. A., and Matta, K. L.**, Synthetic mucin fragments: Benzyl 2-acetamido-6-*O*-(2-acetamido-2-deoxy-β-D-glucopyranosyl)-2-deoxy-3-*O*-β-D-galactopyranosyl-α-D-galactopyranoside and benzyl 2-acetamido-6-*O*-(2-acetamido-2-deoxy-β-D-glucopyranosyl)-3-*O*-[6-*O*-(2-acetamido-2-deoxy-β-D-glucopyranosyl)-β-D-galactopyranosyl]-2-deoxy-α-D-galactopyranoside, *Carbohydr. Res.*, 126, 115, 1984.

386. **Piskorz, C. F., Abbas, S. A., and Matta, K. L.**, Synthesis of benzyl 2-acetamido-2-deoxy-3-*O*-β-D-fucopyranosyl-α-D-galactopyranoside and benzyl 2-acetamido-6-*O*-(2-acetamido-2-deoxy-β-D-galactopyranosyl)-2-deoxy-3-*O*-β-D-fucopyranosyl-α-D-galactopyranoside, *Carbohydr. Res.*, 131, 257, 1984.

387. **Rana, S. S. and Matta, K. L.**, A practical synthesis of 2-acetamido-2-deoxy-3,4-di-*O*-β-D-galactopyranosyl-D-galactopyranose, *Carbohydr. Res.*, 116, 71, 1983.

388. **Paulsen, H., Jacquinet, J. C., and Rust, W.**, Synthese von Oligosaccharid-Determinanten mit Amidspacer vom Typ des T-Antigens, *Carbohydr. Res.*, 104, 195, 1982.

389. **Jacquinet, J. C. and Paulsen, H.**, Synthese von Oligosaccharid-Determinanten des T-Antigens mit Amid-Spacer zur Darstellung synthetischer Antigene, *Tetrahedron Lett.*, 22, 1387, 1981.

390. **Paulsen, H., Paal, H., and Schultz, M.**, Syntheseblock β-D-Gal(1→3)-D-GalNAc zur selektiv-simultanen Anknüpfung an Peptide zu *O*-Glycopeptiden, *Tetrahedron Lett.*, 24, 1759, 1983.

391. **Collins, P. M. and Munasinghe, V. R. N.**, The photochemistry of carbohydrate derivatives. Part 5. Synthesis of methyl 2,3-di-*O*-(β-D-glucopyranosyl)-α-L-fucopyranoside and methyl 2,3-di-*O*-(β-D-galactopyranosyl)-α-L-fucopyranoside using photolabile *O*-(2-nitrobenzylidene) acetals as temporary blocking groups, *J. Chem. Soc. Perkin I*, 921, 1983.

392. **Paulsen, H. and Lorentzen, J. P.**, Synthese der Trisaccharid-Determinanten des Enterobacterial Common Antigens, *Carbohydr. Res.*, 150, 63, 1986.

393. **Paulsen, H. and Lorentzen, J. P.**, Synthese der immunologisch essentiellen Saccharidsequenz des "Enterobacterial Common Antigens", *Angew. Chem.*, 97, 791, 1985.

394. **Suami, T., Otake, T., Kato, N., Nishimura, T., and Ikeda, T.**, Synthesis of gentianose, *Carbohydr. Res.*, 21, 451, 1972.

395. **Suami, T., Otake, T., Nishimura, T., and Ikeda, T.**, Synthesis of raffinose and an isomer, *Carbohydr. Res.*, 26, 234, 1973.

396. **Lemieux, R. U., Wong, T. C., and Thøgersen, H.**, The synthesis and conformational properties of the diastereoisomeric βD-Gal(1→4)βD-GlcNAc (1→6)6-*C*-CH₃-D-Gal trisaccharides, *Can. J. Chem.*, 60, 81, 1982.

397. **Kosma, P., Schulz, G., and Unger, F. M.**, Syntheses of repeating units of *Escherichia coli* capsular polysaccharides containing D-ribose and 3-deoxy-D-*manno*-2-octulosonic acid (KDO), *Carbohydr. Res.*, 132, 261, 1984.

398. **Thiem, J. and Ossowski, P.**, Synthesen und Reaktionen von Methyl-digilanidobiosid, *Chem. Ber.*, 114, 733, 1981.

399. **Thiem, J., Ossowski, P., and Schwentner, J.**, Synthese neuer Mono- und Disaccharidglycale zum Aufbau von Oligodesoxyoligosacchariden, *Chem. Ber.*, 113, 955, 1980.

400. **Kováč, P. and Glaudemans, C. P. J.**, Synthesis of methyl glycosides of β-(1→6)-linked D-galactobiose, galactotriose, and galactotetraose having a 3-deoxy-3-fluoro-β-D-galactopyranoside end-residue, *Carbohydr. Res.*, 140, 289, 1985.

401. **Kováč, P., Glaudemans, C. P. J., Guo, W., and Wong, T. C.**, Synthesis of methyl *O*-(3-deoxy-3-fluoro-β-D-galactopyranosyl)-(1→6)-*O*-β-D-galactopyranosyl-(1→6)-3-deoxy-3-fluoro-β-D-galactopyranoside and related N. M. R. studies, *Carbohydr. Res.*, 140, 299, 1985.

402. **Wiesner, K., Tsai, T. Y. R., and Jin, H.**, On cardioactive steroids. XVI. Stereoselective β-glycosylation of digitoxose: the synthesis of digitoxin, *Helv. Chim. Acta*, 68, 300, 1985.

403. **Thiem, J., Ossowski, P., and Ellermann, U.**, Über die Einführung von Digitoxose-Einheiten und gezielte β-Verknüpfungen bei synthesen von 2′-Desoxydisacchariden, *Liebigs Ann. Chem.*, 2228, 1981.

404. **Thiem, J. and Gerken, M.**, Synthesis of the E-D-C trisaccharide unit of aureolic acid cytostatics, *J. Org. Chem.*, 50, 954, 1985.

405. **Monneret, C., Martin, A., and Pais, M.**, Synthesis of the oligosaccharide moieties of musettamycin and marcellomycin, new antitumor antibiotics, *Tetrahedron Lett.*, 27, 575, 1986.

406. **Martin, A., Pais, M., and Monneret, C.**, Synthesis of a trisaccharide related to the antitumour antibiotic, aclacinomycin A, *J. Chem. Soc. Chem. Commun.*, 305, 1983.

INDEX

β-D-Glcp-(1→3)-D-Gal-(4←1)-β-D-Glcp, 209
β-D-Glcp-(1→2)-β-D-Galp-(1→6)-D-Gal, 201
α-D-Glcp-(1→2)-α-D-Galp-(1→3)-D-Glc, 48
α-D-Glcp-(1→6)-D-Gal-(3←1)-α-L-Rhap, 208
β-D-Glcp-(1→3)-D-Gal-(2←1)-α-L-Rhap, 210
α,β-D-Glcp-(1→3)-D-GlcA-(4←1)-α,β-D-Glcp, 92
α-D-Glcp-(1→2)-D-Glc-(4←1)-α-D-Glcp, 77
α-D-Glcp-(1→2)-D-Glc-(6←1)-α-D-Glcp, 77
α-D-Glcp-(1→3)-D-Glc-(6←1)-α-D-Glcp, 78
α-D-Glcp-(1→3)-D-Glc-(6←1)-β-D-Glcp, 79
α-D-Glcp-(1→4)-D-Glc-(6←1)-α-D-Glcp, 80
α-D-Glcp-(1→4)-D-Glc-(6←1)-β-D-Glcp, 81
α-D-Glcp-(1→6)-D-Glc-(2←1)-β-D-Glcp, 83
α-D-Glcp-(1→6)-D-Glc-(3←1)-β-D-Glcp, 83
β-D-Glcp-(1→2)-D-Glc-(3←1)-β-D-Glcp, 84
β-D-Glcp-(1→2)-D-Glc-(4←1)-α-D-Glcp, 81
β-D-Glcp-(1→2)-D-Glc-(6←1)-β-D-Glcp, 85
β-D-Glcp-(1→3)-D-Glc-(6←1)-β-D-Glcp, 87
β-D-Glcp-(1→4)-D-Glc-(2←1)-β-D-GlcpA, 87
β-D-Glcp-(1→4)-D-Glc-(6←1)-β-D-GlcpA, 87
β-D-Glcp-(1→6)-α-D-Glcp-(1→2)-β-D-Fruf, 233
β-D-Glcp-(1→2)-β-D-Glcp-(1→4)-D-Gal, 176
β-D-Glcp-(1→4)-α-D-Glcp-(1→6)-D-Gal, 175
β-D-Glcp-(1→4)-β-D-Glcp-(1→6)-D-Gal, 176
α-D-Glcp-(1→2)-α-D-Glcp-(1→2)-D-Glc, 19
α-D-Glcp-(1→2)-α-D-Glcp-(1→6)-D-Glc, 24
α-D-Glcp-(1→2)-β-D-Glcp-(1→2)-D-Glc, 29
α-D-Glcp-(1→2)-β-D-Glcp-(1→6)-D-Glc, 36
α-D-Glcp-(1→3)-α-D-Glcp-(1→6)-D-Glc, 24
α-D-Glcp-(1→3)-β-D-Glcp-(1→6)-D-Glc, 36
α-D-Glcp-(1→4)-α-D-Glcp-(1→6)-D-Glc, 25
α-D-Glcp-(1→4)-β-D-Glcp-(1→6)-D-Glc, 37
α-D-Glcp-(1→6)-α-D-Glcp-(1→4)-D-Glc, 20
α-D-Glcp-(1→6)-α-D-Glcp-(1→6)-D-Glc, 26
α-D-Glcp-(1→6)-β-D-Glcp-(1→6)-D-Glc, 38
β-D-Glcp-(1→2)-α-D-Glcp-(1→6)-D-Glc, 27
β-D-Glcp-(1→2)-β-D-Glcp-(1→2)-D-Glc, 30
β-D-Glcp-(1→2)-β-D-Glcp-(1→6)-D-Glc, 39
β-D-Glcp-(1→3)-α-D-Glcp-(1→6)-D-Glc, 27
β-D-Glcp-(1→3)-β-D-Glcp-(1→4)-D-Glc, 31
β-D-Glcp-(1→3)-β-D-Glcp-(1→6)-D-Glc, 40
β-D-Glcp-(1→4)-α-D-Glcp-(1→6)-D-Glc, 28
β-D-Glcp-(1→4)-β-D-Glcp-(1→3)-D-Glc, 30
β-D-Glcp-(1→4)-β-D-Glcp-(1→4)-D-Glc, 32
β-D-Glcp-(1→4)-β-D-Glcp-(1→6)-D-Glc, 41
β-D-Glcp-(1→6)-α-D-Glcp-(1→4)-D-Glc, 21
β-D-Glcp-(1→6)-α-D-Glcp-(1→6)-D-Glc, 28
β-D-Glcp-(1→6)-β-D-Glcp-(1→6)-D-Glc, 42
β-D-Glcp-(1→4)-β-D-GlcA-(1→5)-D-GlcA, 91
β-D-Glcp-(1→4)-β-D-Glcp-(1→4)-D-GlcNAc, 94
α-D-Glcp-(1→4)-α-D-Glcp-(1→1)-α-D-Glcp, 17
α-D-Glcp-(1→4)-β-D-Glcp-(1→1)-α-D-Glcp, 18
α-D-Glcp-(1→6)-α-D-Glcp-(1→1)-α-D-Glcp, 17
β-D-Glcp-(1→4)-β-D-GlcpNAc-(1→4)-D-GlcNAc,
   99
β-D-Glcp-(1→4)-β-D-GlcpN-(1→6)-D-Gal, 180
β-D-Glcp-(1→6)-α-D-Manp-(1→2)-D-Glc, 45
β-D-GlcpNAc-(1→3)-D-Gal-(4←1)-β-D-GlcpNAc,
   211
β-D-GlcpNAc-(1→3)-D-Gal-(6←1)-β-D-GlcpNAc,
   212

α-D-GlcpNAc-(1→3)-D-Gal-(4←1)-β-D-Manp, 211
β-D-GlcpNAc-(1→3)-D-Gal-(4←1)-β-D-Manp, 212
β-D-GlcpNAc-(1→3)-D-GalNAc-(6←1)-α-D-
   GlcpNAc, 229
β-D-GlcpNAc-(1→3)-β-D-Galp-(1→3)-D-GalNAc,
   224
β-D-GlcpNAc-(1→3)-β-D-Galp-(1→4)-D-Glc, 53
β-D-GlcpNAc-(1→4)-β-D-Galp-(1→4)-D-Glc, 54
β-D-GlcpNAc-(1→6)-β-D-Galp-(1→4)-D-Glc, 55
β-D-GlcpNAc-(1→6)-β-D-Galp-(1→4)-D-GlcNAc,
   115
β-D-GlcpNAc-(1→4)-β-D-GlcpNAc-(1→6)-D-Glc,
   44
β-D-GlcpNAc-(1→3)-β-D-GlcpNAc-(1→3)-D-
   GlcNAc, 99
β-D-GlcpNAc-(1→3)-β-D-GlcpNAc-(1→6)-D-
   GlcNAc, 103
β-D-GlcpNAc-(1→4)-β-D-GlcpNAc-(1→6)-D-
   GlcNAc, 103
β-D-GlcpNAc-(1→6)-β-D-GlcpNAc-(1→4)-D-
   GlcNAc, 100
β-D-GlcpNAc-(1→2)-D-Man-(6←1)-β-D-GlcpNAc,
   145
β-D-GlcpNAc-(1→3)-D-Man-(6←1)-β-D-GlcpNAc,
   146
β-D-GlcpNAc-(1→4)-D-Man-(4←1)-β-D-GlcpNAc,
   144
β-D-GlcpNAc-(1→2)-α-D-Manp-(1→6)-D-Man, 139
β-D-GlcpNAc-(1→2)-α-L-Rhap-(1→2)-L-Rha, 157
α-D-GlcpN-(1→3)-D-Gal-(4←1)-β-D-Manp, 211
β-D-GlcpN-(1→3)-D-Gal-(4←1)-β-D-Manp, 212
β-D-GlcpN-(1→3)-D-GalN-(6←1)-α-D-GlcpN, 229
α-D-GlcpN-(1→3)-β-D-Galp-(1→4)-D-Glc, 51
α-D-GlcpN-(1→4)-β-D-Galp-(1→4)-D-Glc, 52
β-D-GlcpN-(1→3)-β-D-Galp-(1→4)-D-Glc, 53
β-D-GlcpN-(1→4)-β-D-Galp-(1→4)-D-Glc, 54
α-D-GlcpN-(1→4)-β-D-Galp-(1→4)-D-GlcN, 114
β-D-GlcpN-(1→6)-β-D-Galp-(1→4)-D-GlcN, 115
α-D-GlcpN-(1→3)-α-D-Galp-(1→4)-L-Rha, 166
α-D-GlcpN-(1→3)-β-D-Galp-(1→4)-L-Rha, 168
β-D-GlcpN-(1→3)-α-D-Galp-(1→4)-L-Rha, 167
β-D-GlcpN-(1→3)-β-D-Galp-(1→4)-L-Rha, 168
α-D-GlcpN-(1→3)-D-GlcN-(4←1)-β-D-Ribf, 121
α-D-GlcpN-(1→4)-β-D-GlcpA-(1→4)-D-GlcN, 97
α-D-GlcpN-(1→6)-α-D-GlcpN-(1→6)-D-GlcN, 93,
   98
α-GlcpN-(1→3)-α-D-GlcpN-(1→3)-D-GlcN, 97
β-D-GlcpN-(1→3)-β-D-GlcpN-(1→3)-D-GlcN, 99
β-D-GlcpN-(1→6)-α-D-GlcpN-(1→6)-D-GlcN, 98
β-D-GlcpN-(1→2)-D-Man-(4←1)-β-D-GlcpN, 144
β-D-GlcpN-(1→2)-D-Man-(6←1)-β-D-GlcpN, 145
β-D-GlcpN-(1→3)-D-Man-(6←1)-β-D-GlcpN, 146
β-D-GlcpN-(1→4)-β-D-Manp-(1→4)-D-GlcN, 106
β-D-GlcpN-(1→2)-α-D-Manp-(1→3)-D-Man, 138
β-D-GlcpN-(1→2)-α-D-Manp-(1→6)-D-Man, 139
β-D-GlcpN-(1→6)-α-D-Manp-(1→6)-D-Man, 140
β-D-GlcpN-(1→2)-α-L-Rhap-(1→2)-L-Rha, 157
α-D-GlcpNSO₃ (3,6-OSO₃)-(1→4)-α-L-IdopA
   (2-OSO₃)-(1→4)-D-GlcNSO₃ (6-OSO₃), 93
β-D-Glcp-(1→3)-L-Rha-(4←1)-β-D-Glcp, 173
α-D-Glcp-(1→3)-L-Rha-(2←1)-β-D-GlcpN, 170

α-D-Glcp-(1→3)-L-Rha-(2←1)-β-D-GlcpNAc, 170

α-D-Glcp-(1→3)-L-Rha-(4←1)-β-D-ManpN, 171

β-D-Glcp-(1→3)-L-Rha-(4←1)-β-D-ManpN, 173

α-D-Glcp-(1→3)-L-Rha-(4←1)-β-D-ManpNAc, 171

β-D-Glcp-(1→4)-α-L-Rhap-(1→3)-D-Gal, 186

β-D-Glcp-(1←4)-α-L-Rhap-(1→4)-D-Gal, 193

β-D-Glcp-(1→4)-α-L-Rhap-(1→6)-D-Gal, 194

α-D-Glcp-(1→3)-α-L-Rhap-(1→3)-D-ManN, 152

α-D-Glcp-(1→3)-α-L-Rhap-(1→3)-D-ManNAc, 152

α-D-Glcp-(1→3)-α-L-Rhap-(1→2)-L-Rha, 155

α-D-Glcp-(1→3)-α-L-Rhap-(1→3)-L-Rha, 160

β-D-Glcp-(1→3)-α-L-Rhap-(1→2)-L-Rha, 156

β-D-Glcp-(1→3)-α-L-Rhap-(1→3)-L-Rha, 161

α-D-Glcp-(1→2)-L-Rha-(3←1)-α-L-Rhap, 169

α-D-Glcp-(1→3)-L-Rha-(2←1)-α-L-Rhap, 171

β-D-Glcp-(1→2)-L-Rha-(3←1)-α-L-Rhap, 172

β-D-Glcp-(1→3)-D-Xyl-(5←1)-β-D-Glcp, 16

K

β-KDO-(2→3)-β-D-GlcpNAc-(1→6)-D-GlcN, 105

α-KDO-(2→3)-β-D-GlcpNAc-(1→6)-D-GlcNAc, 104

α-KDO-(2→3)-β-D-GlcpN-(1→6)-D-GlcN, 104

α-KDO-(2→6)-β-D-GlcpN-(1→6)-D-GlcN, 105

α-KDO-(2→4)-α-KDO-(2→6)-D-GlcN, 94

α-KDO-(2→2)-β-D-Ribf-(1→2)-D-Rib, 2

α-KDO-(2→3)-D-Rib-(2←1)-α-D-Ribf, 2, 3

L

α-D-Lyxp-(1→3)-D-Man-(6←1)-α-D-Manp, 143

α-D-Lyxp-(1→6)-D-Man-(3←1)-α-D-Manp, 143

M

β-D-Manp-(1→4)-α-D-Galp-(1→4)-L-Rha, 167

α-D-Manp-(1→6)-β-D-GlcpNAc-(1→4)-D-GlcNAc, 101

β-D-Manp-(1→4)-β-D-GlcpNAc-(1→4)-D-GlcNAc, 102

α-D-Manp-(1→4)-β-D-GlcpN-(1→4)-D-GlcN, 100

β-D-Manp-(1→4)-β-D-GlcpN-(1→4)-D-GlcN, 102

β-D-Manp-(1→4)-β-D-GlcpN-(1→6)-D-GlcN, 104

α-D-Manp-(1→2)-D-Glc-(6←1)-α-L-Rhap, 88

α-D-Manp-(1→2)-D-Man-(4←1)-α-D-Manp, 146

α-D-Manp-(1→2)-D-Man-(6←1)-α-D-Manp, 147

α-D-Manp-(1→3)-D-Man-(6←1)-α-D-Manp, 148

α-D-Manp-(1→2)-α-D-Manp-(1→3)-D-Gal, 183

α-D-Manp-(1→2)-β-D-Manp-(1→3)-D-Gal, 184

α-D-Manp-(1→2)-α-D-Manp-(1→3)-D-Glc, 46

α-D-Manp-(1→3)-β-D-Manp-(1→4)-D-GlcN, 107

α-D-Manp-(1→6)-β-D-Manp-(1→4)-D-GlcN, 108

α-D-Manp-(1→2)-β-D-Manp-(1→4)-D-GlcNAc, 106

α-D-Manp-(1→3)-β-D-Manp-(1→4)-D-GlcNAc, 107

α-D-Manp-(1→6)-β-D-Manp-(1→4)-D-GlcNAc, 108

α-D-Manp-(1→2)-α-D-Manp-(1→2)-D-Man, 135

α-D-Manp-(1→2)-α-D-Manp-(1→3)-D-Man, 138

α-D-Manp-(1→2)-α-D-Manp-(1→6)-D-Man, 140

α-D-Manp-(1→2)-β-D-Manp-(1→2)-D-Man, 141

α-D-Manp-(1→2)-β-D-Manp-(1→3)-D-Man, 141

α-D-Manp-(1→2)-β-D-Manp-(1→6)-D-Man, 142

α-D-Manp-(1→3)-α-D-Manp-(1→3)-D-Man, 139

β-D-Manp-(1→2)-α-D-Manp-(1→2)-D-Man, 137

β-D-ManpNAcA-(1→4)-α-D-GlcpNAc-(1→3)-D-Fucp4NAc, 237

β-D-ManpNAc-(1→3)-α-L-FucpNAc-(1→3)-D-GalNAc, 227

β-D-ManpNA-(1→4)-α-D-GlcpN-(1→3)-D-Fucp4N, 237

β-D-ManpN-(1→3)-α-L-FucpN-(1→3)-D-GalN, 227

β-D-ManpN-(1→3)-β-L-FucpN-(1→3)-D-GalN, 227

α-D-Manp-(1→4)-α-L-Rhap-(1→3)-D-Gal, 188

α-D-Manp-(1→4)-β-L-Rhap-(1→3)-D-Gal, 197

β-D-Manp-(1→4)-α-L-Rhap-(1→3)-D-Gal, 190

β-D-Manp-(1→4)-α-L-Rhap-(1→4)-D-Gal, 193

β-D-Manp-(1→4)-β-L-Rhap-(1→3)-D-Gal, 198

α-D-Manp-(1→4)-α-L-Rhap-(1→3)-D-Glc, 47

4-O-Me-α-D-GlcpA-(1→2)-β-D-Xylp-(1→4)-D-Xyl, 11

4-O-Me-β-D-GlcpA-(1→2)-β-D-Xylp-(1→4)-D-Xyl, 11

4-O-Me-α-D-GlcpA-(1→2)-D-Xyl-(4←1)-β-D-Xylp, 15

4-O-Me-β-D-GlcpA-(1→2)-D-Xyl-(4←1)-β-D-Xylp, 16

3,6-di-O-Me-β-D-Glcp-(1→4)-2,3-di-O-Me-α-L-Rhap-(1→2)-3-O-Me-L-Rha, 156

3,6-di-O-Me-β-D-Glcp-(1→4)-2,3-di-O-Me-β-L-Rhap-(1→2)-3-O-Me-L-Rha, 164

N

α-Neu5Ac-(2→6)-β-D-Galp-(1→4)-D-GlcN, 120

β-Neu5Ac-(2→6)-β-D-Galp-(1→4)-D-GlcN, 120

α-Neu5Ac-(2→6)-β-D-Galp-(1→4)-D-GlcNAc, 120

β-Neu5Ac-(2→6)-β-D-Galp-(1→4)-D-GlcNAc, 120

α-D-Neup5Ac-(2→3)-β-D-Galp-(1→4)-D-Glc, 68

α-D-Neup5Ac-(2→6)-β-D-Galp-(1→4)-D-Glc, 69

β-D-Neup5Ac-(2→3)-β-D-Galp-(1→4)-D-Glc, 70

β-D-Neup5Ac-(2→6)-β-D-Galp-(1→4)-D-Glc, 71

β-D-Neup5Ac-(2→8)-β-D-Neup5Ac-(2→6)-D-Glc, 76

R

α-L-Rhap-(1→3)-D-Gal-(4←1)-α-L-Rhap, 213

α-L-Rhap-(1→3)-D-GlcA-(4←1)-α-L-Rhap, 92

α-L-Rhap-(1→4)-β-D-Glcp-(1→6)-D-Glc, 43

α-L-Rhap-(1→3)-β-D-GlcpNAc-(1→2)-L-Rha, 153

α-L-Rhap-(1→2)-α-L-Rhap-(1→6)-D-Gal, 195

α-L-Rhap-(1→3)-α-L-Rhap-(1→6)-D-Gal, 196

α-L-Rhap-(1→4)-α-L-Rhap-(1→6)-D-Gal, 197

α-L-Rhap-(1→3)-α-L-Rhap-(1→6)-D-Glc, 47

α-L-Rhap-(1→3)-α-L-Rhap-(1→3)-D-GlcNAc, 109

α-L-Rhap-(1→4)-α-L-Rhap-(1→3)-D-GlcNAc, 110

α-L-Rhap-(1→2)-α-L-Rhap-(1→2)-L-Rha, 158

α-L-Rhap-(1→2)-α-L-Rhap-(1→3)-L-Rha, 162

α-L-Rhap-(1→3)-α-L-Rhap-(1→2)-L-Rha, 159

α-L-Rhap-(1→3)-α-L-Rhap-(1→3)-L-Rha, 163

α-L-Rhap-(1→4)-α-L-Rhap-(1→2)-L-Rha, 160

T

X

Printed and bound by CPI Group (UK) Ltd, Croydon, CR0 4YY

22/10/2024

01777638-0010